CAMBRIDGE COUNTY GEOGRAPHIES

General Editor: F. H. H. GUILLEMARD, M.A., M.D.

T0352318

STAFFORDSHIRE

Cambridge County Geographies

STAFFORDSHIRE

by

W. BERNARD SMITH, B.Sc.

Senior Science Master at Denstone College

With Maps, Diagrams and Illustrations

Cambridge :

at the University Press

1915

CAMBRIDGE UNIVERSITY PRESS

Cambridge, New York, Melbourne, Madrid, Cape Town,
Singapore, São Paulo, Delhi, Mexico City

Cambridge University Press
The Edinburgh Building, Cambridge CB2 8RU, UK

Published in the United States of America by Cambridge University Press, New York

www.cambridge.org
Information on this title: www.cambridge.org/9781107663527

© Cambridge University Press 1915

This publication is in copyright. Subject to statutory exception
and to the provisions of relevant collective licensing agreements,
no reproduction of any part may take place without the written
permission of Cambridge University Press.

First published 1915
First paperback edition 2013

A catalogue record for this publication is available from the British Library

ISBN 978-1-107-66352-7 Paperback

This publication reproduces the text of the original edition of the
Cambridge County Geographies. The content of this publication has
not been updated. Cambridge University Press has no responsibility
for the accuracy of the geographical guidance or other information
contained in this publication, and does not guarantee that such
content is, or will remain, accurate.

PREFACE

THE author wishes to acknowledge the information he has obtained from the *Annual Reports of the North Staffordshire Field Club*, Mr C. Masefield's *Staffordshire*, the *Victoria History of Staffordshire* and *Kelly's Directory*.

He is indebted to the Rev. F. A. Hibbert, Headmaster of Denstone, for the chapter on Monasticism ; to Messrs F. T. Howard (H.M. Chief Inspector of Schools) and C. Lynam and some of his colleagues on the Denstone staff for valued criticism ; and to members of his family for assistance in the preparation of the manuscript for the press.

Denstone College, March 1915.

CONTENTS

CONTENTS

ILLUSTRATIONS

MAPS

The illustrations on pp. 19 and 31 are from photographs by the author; those on pp. 9, 11, 12, 20, 48, 70, 73, 87, 136, 140, are from photographs by Messrs R. and R. Bull, Ashbourne; those on pp. 22, 59, 61, 83, 97, 108, 137, 148 are from photographs by Messrs H. J. Gover and Co., Hanley; those on pp. 6, 15, 34, 37, 49, 76, 85, 89, 91, 98, 117, 126, are from photographs by Mr A. McCann, Uttoxeter; those on pp. 54, 77, 81, 88, 101, 145, 150 are from photographs by Mr James Gale, Wolverhampton; those on pp. 66 and 146 are from photographs by Messrs J. Valentine and Sons; those on pp. 4, 94, 132 are from photographs by Mr B. Lowndes, Cheadle; that on p. 79 is from a photograph by Messrs Frith.

1. County and Shire. Origin of Stafford-shire.

A reader who glances at the map placed inside the front cover of this book will notice that Staffordshire is a roughly oval tract of land with a boundary that seems to have a haphazard course, except where it follows the windings of a river. No greater contrast between such boundaries and those we find in the divisions of Australia, the Argentine, or the various parts of the United States could be imagined. The difference between the irregular and apparently arbitrary outlines of ancient political divisions and the straight boundaries of modern times is like the contrast between the narrow, picturesquely winding streets of our ancient cities and the broad, straight streets of newly-planned colonial towns. The reasons for the differences are similar. The towns of the Old World are generally the result of natural growth round some castle, mill, or harbour; but modern towns are often planned out on paper before a brick is laid or a corner-post driven in. Modern Boundary Commissioners can use the imaginary lines of latitude and longitude as their limits, but older countries are generally bounded by natural frontiers of hills or river courses, or are the results of divisions by treaty. Sometimes the union of many smaller units, themselves of irregular shape, has built up a larger division. Staffordshire is an example of this construction.

While the precise date of origin of county divisions in the middle of England is still a historical problem, we shall probably not be far wrong if we say that Staffordshire as a county is about a thousand years old. The Danish tide of conquest washed out many ancient boundaries, and new divisions were required when the ebb came. King Alfred did much to check the progress of the invaders, and his son, Edward the Elder, continued his father's efforts. Edward was helped by his wise and warlike sister, Ethelfleda, or Aethelflaed, the "Lady of the Mercians." Ethelfleda fought well against the Danes in Mercia—as this part of England was called—and built or restored forts in the districts she recovered. In the Anglo-Saxon Chronicle we read: "A.D. 913...This year by the permission of God went Ethelfleda, Lady of Mercia, with all the Mercians to Tamworth; and built the fort there in the fore-part of the summer; and before Lammas that at Stafford: in the next year that at Eddesbury."

Stafford (i.e. the ford at the *Staeth* or jetty) must have been more important than Lichfield, in spite of the bishop (and, for some time, an archbishop) having his seat at the latter place, otherwise it would not have been chosen as the county town. It was the chief town of five adjacent "hundreds" which were placed under one administration at about this time, and called the *Scire* or *Shire* of Staeth.

The term *shire* signified that it was the part *shorn* off, or cut off to form its *share*. These hundreds had irregular boundaries, and the outer limits of the new Staffordshire were therefore irregular too.

The other shires of Middle England, especially those whose names are the same as those of their chief towns, were probably formed in the same way. In the south the land was already divided into kingdoms such as Sussex, the land of the South Saxons, and Kent, the kingdom of the Cantii. When these kingdoms came under the general rule of one overlord, earls—or, as the Normans called them, counts—ruled over them. The English term *Earldom* dropped out of use after the Norman Conquest, and *Comté*, or county, the conquerors' word, persisted as the official name. Historically then, it is incorrect to call such counties as Middlesex, Sussex, or Kent by the name of shires, or to call Warwickshire, Derbyshire, or Staffordshire counties, but the latter word is nowadays used for all.

Many alterations have taken place since that time, both in the number and shape of the counties. Some counties such as Norhamshire, Richmondshire, Hexhamshire and Winchelcombeshire have been merged into others, while in recent years simplification of boundaries has frequently taken place in order that local government may be carried on more easily. The latest alteration in our own county is the addition of Handsworth to the city of Birmingham in 1910 and its transference from Staffordshire to Warwickshire.

2. General Characteristics.

Staffordshire lies at the north-western border of the group of the midland counties, on the direct line between London and Chester, at the point where the Pennine Chain falls gradually to lose itself in the central plain

Dimmings Dale, near Alton

of England. Its main physical character, except in the north, where the long ridges and deep valleys of the "Moorlands" run north-west and south-east to the Trent valley, is that of level or gently undulating land, of no great height. In early times much of this, no doubt, was covered with forest, though that day has long passed and the county nowadays, owing to its large population, is thickly meshed with a network of roads, railways, and canals.

A very large proportion of its area—some four-fifths—is arable, and the county was at one time famous for its barley. Owing, however, to the nearness of large centres of population the farmers of Staffordshire have found that dairy-farming is more profitable, and nowadays in many districts almost the whole of the land is under grass. The character of the soil varies with the underlying geological formation : the New Red Sandstone produces a warm and light soil, but of a poor character chemically ; on the other hand the soil disintegrated from the red Keuper marl is heavy and rather cold, but its chemical composition and moisture-retaining properties render it very suitable for grass. The hilly districts of the north-east are also suitable for grass.

Cut off from the sea by the surrounding counties and its largest river only becoming navigable for barges as it leaves the county, Staffordshire for centuries did not share in the general progress of England to the same extent as the maritime and home counties. But coal has altered all this, and has made Staffordshire third in point of importance of all the counties in England for manufactures. In the north-west we have the densely-crowded Potteries—the chief seat of the earthenware manufacture in the kingdom. In the south is the great Dudley coalfield, with coal of remarkably good quality, and the unlovely " Black Country," where, with Wolverhampton and Walsall as centres, iron is very largely manufactured. In the extreme east is Burton-on-Trent with its vast brewing industry. Favoured by its position on the main line of traffic and by facilities for reaching the chief markets both at home and abroad Staffordshire has developed very rapidly, and as might be expected, its population has increased enormously.

Hanley, a typical pottery town

3. Size. Shape. Boundaries.

Staffordshire is seventeenth in size among the fifty-two counties of England and Wales. Its present area, 741,298 acres, or about 1159 square miles, is rather smaller than its ancient dimensions, for during the nineteenth century administrative changes were made which gave it 1594 acres

formerly in Derbyshire and Warwickshire, but took away
3969 acres for the benefit of Worcestershire, and in 1910
Handsworth was taken from Staffordshire and added to
Warwickshire. Its greatest length, from north to south,
is about 59 miles, and its greatest breadth 37 miles.

The county is roughly oval, resembling a lozenge-
shaped inland lake with many promontories and indenta-
tions, and even an island in the south—a little bit of
Worcestershire separated from its mainland. This isolated
portion of Worcestershire was owned by one of the
monastic houses of that county and was placed under the
same jurisdiction as the rest of the property.

Let us trace the county boundary in a westerly
direction beginning at its northern limit on the side of
Axe Edge, where, at Three Shires Head, the counties of
Stafford, Chester, and Derby meet. For a few miles it
follows the windings of the river Dane. Near Bosley it
turns abruptly south-west and, leaving Congleton to the
north-west, climbs Congleton Edge, running along the top
and over Mow Cop (977 feet). After this the boundary
travels, with no obvious natural feature to guide it, nearly
as far as Woore, where Cheshire gives place to Shropshire
as neighbour. Then the little river Tern is met with,
which takes the boundary nearly to Market Drayton.
Here the line bends abruptly east, round Blore Heath, till
turned south again by Bishop's Wood. The course is
now irregular though still southerly, making use of the
river Mees for about four miles. It just misses Newport,
after which it leaves the low land containing Aqualate
Mere and runs uphill, crossing Watling Street, and passing

hard by Boscobel and Kingswood and over the high land to the west of Wolverhampton. Thence it drops down again to the basin of the Stour, enclosing Abbot's Castle Hill in Staffordshire. After rising once more it runs along the watershed between the Stour and the Severn and suddenly turns south-east. Upper Arley and a part of the Wyre Forest were in Staffordshire until 1895, when this curious extension, shown uncoloured on the map, was transferred to Worcestershire.

After rising to the high land on the east of the Stour the boundary turns abruptly north until it again meets that river, passing close by the town of Stourbridge. The line thence runs very irregularly north-east, enclosing most of the Black Country, and, leaving Birmingham to the south and west, follows the Roman Ryknield Street as far as Watford Gap. Here there is another change of direction eastward which lasts until the Tame is met. The boundary is made by this river as far as Tamworth, which lies partly in Warwickshire; then the north-easterly trend is resumed. Near the crossing of the Mease Leicestershire is the next county, but soon gives place to Derbyshire. The Trent lies between Derbyshire and Staffordshire till about two miles below Burton-on-Trent. Here the boundary leaves the river for a short distance and runs along the high land to the east of Burton, thus enclosing the whole of the town. Then the river Dove forms the boundary, except for a slight aberration at Rocester, almost to our starting point at Axe Edge. In this latter part of its course the boundary line has run gradually uphill from the low-lying tracts near Burton right to the Weaver Hills, Dove Dale, and the Peak District.

As we have already seen, the irregularities of the boundary are due to the building-up of the county by the grouping of "hundreds." The hundreds are composed of

The Dove at Hanging Bridge

(*The boundary between Staffordshire and Derbyshire*)

parishes and estates, and these have boundaries depending upon minor physical features or local conditions at the time of formation. There is a tendency in modern times for county boundaries to be modified in order to simplify local government.

4. Surface and General Features.

The map placed inside the front cover shows at a glance the general character of the surface of the county.   The dark brown colouration in the north indicates that portion of the southern end of the Pennine Chain which is more than nine hundred feet above sea-level.   This is the "Moorlands," or stone-wall district of Staffordshire, and if we look now at the map at the end of the book we shall see that this region consists almost entirely of limestone.   Camden, who wrote in Latin a geographical work called *Britannia*, of which an English translation was published in 1611, well says of this region, "The North part riseth up and swelleth somewhat mountainous, with moores and hilles, but of no great bignesse, which beginning here, runs like as *Apennine* doth in Italie, through the middest of England with a continued ridge, rising more and more with divers tops and cliffs one after another even as far as to Scotland, although often-times they change their name.   For heere they are called Mooreland, after a while the *Peak*, *Blackstone Edge*, then *Craven*, anon as they go further *Stanmore*, and at length being parted diversly as it were into hornes, *Cheviot*.   This *Mooreland*, so called for that it riseth higher into hils and mountaines, is a small country verily; so hard, so comfortlesse, bare, and cold, that it keepeth snow lying upon it a long while: in so much as that of a little country village named Wotton lying here under Weverhill [Wootton is two miles north-west of Ellastone] the neighbor inhabitants

Throwley Old Hall

(Shewing the Moorlands or "stone-wall district" of Staffordshire)

have this rime rise in their mouth, as if God, forsooth, had never visited that place.

> ' Wotton under Wever
> Where God came never.'

Yet in so hard a soile it breedeth and feedeth beasts of large bulke, and faire spread."

Dovedale from Reynards Cave

The scenery in this region is characterised by the presence of loose stone walls instead of hedges. One comes across these quite suddenly on reaching the limestone outcrop. The hills are covered with closely-cropped herbage, with clumps of trees here and there all bent towards the north-east by the prevailing winds; for on these high lands there is nothing to break the force of gales, and there is usually a breeze to be found there. The Dove has cut itself a deep gorge through the rocks, the famous Dove Dale, well known to lovers of scenery and to disciples of Izaak Walton.

Less known outside Staffordshire, but no less deserving of a visit, are Glutton and Cotton Dales, and the valleys of the Manifold and Hamps. These two rivers have underground channels for parts of their course. In dry weather the whole of their waters disappear through holes in the river bed, coming to the surface again in the lovely glen called " Paradise," at Ilam. When there is too much water for the subterranean passage the overflow follows an upper, stony course to Ilam.

In the extreme north the Roches provide the geologist with a fine example of an "escarpment." The Roches, Morridge, and the Weaver Hills form a ridge with a steep slope towards the south-west. On the north-east are hills separated by deep valleys which have been cut by the combined action of frost and running water in the limestone. Much of this part of Staffordshire is well over a thousand feet above sea-level. The sharp ridge of the Roches rises above the 1500-foot contour line, the highest point having an altitude of 1657 feet. Merryton Low is 1603 feet high and the Weaver Hills reach 1217 feet at Beacon Stoop.

The narrow Churnet Valley, the western boundary of Moorlands, with its steep and well-wooded sides, was carved out of the Triassic Sandstone, and near Alton a tiny tributary flows down the pretty, wooded gorge of Dimmings Dale. In the north-west, stretching from the

Roches to Biddulph Moor and Cloud End, is a tract of fine moorland country. Round about the Potteries the scenery is far from tame, and near Trentham the country is well wooded.

South of this hilly region we find pasture land varied with parks, the remains of the numerous deer enclosures which formerly occupied much land in the county. This region coincides roughly with the brown patch on the geological map at the back of this volume, and indeed is the result of the presence of the red Keuper marl which that colour indicates. This rock produces a heavy soil which retains moisture even in very dry weather, so is favourable to the growth of grass. Just as the valley of the river Lea is specially suitable for the growth of fruit and vegetables for the London market, so also the marl area of Staffordshire has become to a large extent a dairy district supplying the neighbouring populous areas. The river valleys traversing the brown patch are coloured white, indicating that the rivers have been able to deposit much alluvial matter. Thus from the map we can see that they are flowing slowly through low-lying land. These valleys, as we should expect, are in some places very flat, swampy, and occasionally flooded.

Stretching upwards into this region from the hillier southern portion of the county is Cannock Chase. This is a large district of about 30,000 acres, very sparsely populated, covered chiefly with fern and heather and with a very irregular surface which occasionally rises to about 800 feet above sea-level. It was once a forest, containing many splendid oaks. The soil is of a dark colour and is dry and peaty; beneath it is a subsoil of sandstone and gravel. Its wild scenery is somewhat spoilt by ironstone quarries and by the chimneys and pit-banks of coal mines which have been sunk in it to reach the seams which, lying nearer to the surface farther south, pass underneath the Chase. It is probable that the forest of Cannock Chase was at one time far more extensive, possibly being

Beggar's Oak, Bagot's Park

connected with Needwood Forest, west of Burton, which is still a well-wooded district containing magnificent oaks.

South of Cannock Chase we find, contrasting strangely with it, the ugly "Black Country," so called from the smoke and dirt which at present accompany industrial prosperity in this Coal Age through which the civilised world is passing.

The British Isles and the surrounding seas, showing the "Continental Shelf"

5. Watershed. Rivers and their Courses. Lakes.

Though the waters of the North Sea are very shallow, this is not the case to the westward of our islands, where we find the soundings drop suddenly to the great ocean depths. Our islands are simply those portions of a large continental shelf which are high enough to stand above the water, as shown on p. 16. This shelf was once above the water, when clearly the English Channel and the North Sea did not exist except as low plains, and the Irish Sea was merely a long inland lake.

On looking carefully at a map of Great Britain it is seen that the rivers on the east are usually longer than those flowing to the west. The watershed, as is often the case, does not run along the middle of the country. In Scotland only two rivers on the west coast are over sixty miles long, while on the east side there are five of this length. A glance at a map of England and Wales will show that the "backbone" of each country is not in the median line. The Pennine Chain forms the "divide" from the Cheviot Hills as far as North Stafford-shire, and after this the ridge passes along the hills of Worcestershire with a U-shaped bend eastwards and back caused by the Warwickshire Avon, then along the Cotswolds. After this there is a bend eastwards through the southern counties to the Straits of Dover. This ridge appears to have been continuous with the water-parting which separates the basins of the Schelde, Meuse,

and Rhine on the north from those of the Somme and
the Seine on the south. It will be noticed, however,
that some of the finest scenery is to be met with along
the gorges frequently cut by rivers which, in whatever
direction they began to run when the land they drain was
first raised above the ocean, now have their sources on
one side of a range of mountains and flow in deep valleys
through the ridge to the ocean on the other side.

It seems probable that the eastward-flowing rivers of
Britain, and the westward-flowing streams of Scandinavia,
draining lands round the Baltic Sea and Western Europe
as far as the Schelde, used to join into one great river,
probably the continuation of the Rhine, which ran northwards to empty itself into the Arctic Ocean.

The rivers of Staffordshire form a small part of this
great system : it can now be examined in more detail.
On the topmost border of the map two rivers are
marked, the Dove and the Dane, flowing in different
directions from Axe Edge. By tracing their courses it
will be seen that the Dove joins the Trent, which flows
into the North Sea, while the Dane runs to join the
Weaver and eventually empties itself into the estuary of
the Mersey.

Clearly Axe Edge forms a short length of the main
watershed of England. From it the water-parting runs
southwards to the corner of the Roches, then bends
northwards again almost to the boundary. Thence it
drops down to a height of just over 500 feet at a point
north of Rudyard Lake. The natural line of demarca-
tion of the basins of the Trent and Dane is now so

slightly marked that Dane water is taken by a canal feeder across the water-parting to Rudyard Lake, and so into the Trent Basin. The watershed now rises to Biddulph Moor, reaching a height of over eleven hundred feet.

The Dove rises as a spring just within the border of the county at its extreme northern limit. It has gradually

Source of the Dove

worn for itself in the limestone of the Peak one of the most picturesque gorges in England. Beresford Dale, near Hartington, is a short stretch of great beauty; lower down the river runs between high cliffs supporting rocks carved by frost and water into fantastic shapes, in which caves occur here and there.

At Thorpe the gorge widens out and receives the combined waters of the Hamps and Manifold. These rivers disappear completely in dry weather down "swallows" at Waterfall, several miles higher in their course, and run underground, reappearing in Paradise, an exquisite little valley at Ilam. In the illustration the two bays on the

Entrance to Paradise, Ilam

right are their waters coming to the light of day. In wet weather the swallows cannot take all the water, and the overflow runs like an ordinary stream down the picturesque valley. Many attempts have been made to choke the swallows, but they have met with only partial success.

The Dove receives on its right bank at Rocester the waters of the Churnet. It will be found that this river also rises very near the northern border : one of its head waters is but half a mile from a little tributary of the Dane. The Churnet valley is beautiful in many parts, while its fall, like that of the Dove, is very rapid till within a mile or two of their junction, when their valleys become flat land, marshy in places, subject to floods, and covered with the gravel which the waters have rolled down and deposited in these quieter reaches. The course of the Dove is now very winding till it joins the Trent. The bed of the river has probably zig-zagged across its valley many times, for when waters meet a bend they tend to go straight on, and in time wear for themselves a new channel which cuts off the bend. Some slight obstruction, however, soon gives rise to a new bend, and the process is repeated. The river bed thus travels across its valley by slow degrees, and then returns, lowering the whole a little each time.

The Trent, the longest river of Staffordshire, is another that rises right at the main watershed on Biddulph Moor. For the first part of its course it flows through moorland, then plunges into the murkiness of the Potteries, emerging south of Stoke with the addition of the waters of the Lyme. It is now a larger but much dirtier stream, though it rapidly becomes cleaner as it flows through the pretty, well-wooded country round Trentham. By the time the river has reached a point three miles north-west of Stone it has worn its valley down to the three-hundred-foot level, shown in the physical map as the boundary between the brown and green colouration. It has thus fallen nearly eight hundred feet in a distance, as the crow flies, of but twelve miles, while in the remainder of its long course to the Humber it has only three hundred feet to fall. Its flow is now much slower, and alluvium or silt is deposited. Near Great Haywood the Trent receives the waters of the Sow which, with its tributaries, the Meece and the Penk

Source of the Trent

and their subsidiary streams, has drained western Staffordshire almost to the border. A few miles further down the Blythe joins the Trent, bringing the drainage of the land between the Trent and the Churnet.

The main watershed is thus seen to run southwards just inside the county nearly as far south as Wolverhampton, where it turns south-east, then southwards through the middle of the Black Country, and so out of the county.

The Tame rises near Wednesbury, runs south-east into Warwickshire north of Birmingham, bends northwards, and bounds Staffordshire for a few miles south of Tamworth.   At this town it receives the waters of the Anker, from Warwickshire, and then runs into the county.   After winding through low-lying country it joins the Trent near Croxall.

Water runs off the high moorland of Cannock Chase in all directions.   On the south the streams run either into a large canal reservoir near Norton, the overflow being taken by the Black or Bourne Brook into the Tame, or else into the Saredon Brook which runs eastwards to join the Penk.   The Penk takes the westward drainage, while the northward and eastward runs by streamlets into the Sow or the Trent.   Cannock Chase is thus entirely in the basin of the Trent.   In spite of the Trent receiving the surplus water of nearly the whole county it cannot be used for barges until it enters Derbyshire.

The drainage of the strip of land between the main watershed and the western boundary of the county is divided into two unequal portions by a secondary waterparting which branches off the main line at the Potteries and runs westward.   The water falling to the north-west of the Potteries finds its way into the estuary of the Mersey, while the Severn receives the rest of the surplus surface water of the county.   There are no large rivers, the Tern, Mees (draining Aqualate Mere), and Stour being the most important.

6. Geology and Soil.

By Geology we mean the study of the rocks, and we must at the outset explain that the term *rock* is used by the geologist without any reference to the hardness or compactness of the material to which the name is applied; thus he speaks of loose sand as a rock equally with a hard substance like granite.

Rocks are of two kinds, (1) those laid down mostly under water, (2) those due to the action of fire.

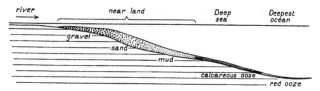

Sectional Diagram showing the various deposits formed at the mouth of a river

The first kind may be compared to sheets of paper one over the other. These sheets are called *beds*, and such beds are usually formed of sand (often containing pebbles), mud or clay, and limestone, or mixtures of these materials. They are laid down as flat or nearly flat sheets, but may afterwards be tilted as the result of movement of the earth's crust, just as we may tilt sheets of paper, folding them into arches and troughs, by pressing them at either end. Again, we may find the tops of the folds so produced worn away as the result of the action of rivers, glaciers, and sea-waves upon them, as we might cut off the tops of the folds of the paper with a pair of shears. This has happened with the ancient beds forming parts of the earth's crust, and we therefore often find them tilted, with the upper parts removed.

The other kinds of rocks are known as igneous rocks; they have been melted under the action of heat and become solid on cooling. When in the molten state they have been poured out at the surface as the lava of volcanoes, or have been forced into other rocks and cooled in the cracks and other places of weakness. Much material is also thrown out of volcanoes as volcanic ash and dust, and is piled up on the sides of the volcano. Such ashy material may be arranged in beds, so that it partakes to some extent of the qualities of the two great rock groups.

The production of beds is of great importance to geologists, for by means of these beds we can classify the rocks according to age. If we take two sheets of paper, and lay one on the top of the other on a table, the upper one has been laid down after the other. Similarly with two beds, the upper is also the newer, and the newer will remain on the top after earth-movements, save in very exceptional cases which need not be regarded by us here, and for general purposes we may regard any bed or set of beds resting on any other in our own country as being the newer bed or set.

The movements which affect beds may occur at different times. One set of beds may be laid down flat, then thrown into folds by movement, the tops of the beds worn off, and another set of beds laid down upon the

	NAMES OF SYSTEMS	SUBDIVISIONS	CHARACTERS OF ROCKS
TERTIARY	Recent Pleistocene	Metal Age Deposits Neolithic „ Palaeolithic „ Glacial „	Superficial Deposits
	Pliocene	Cromer Series Weybourne Crag Chillesford and Norwich Crags Red and Walton Crags Coralline Crag	Sands chiefly
	Miocene	Absent from Britain	
	Eocene	Fluviomarine Beds of Hampshire Bagshot Beds London Clay Oldhaven Beds, Woolwich and Reading Thanet Sands [Groups	Clays and Sands chiefly
SECONDARY	Cretaceous	Chalk Upper Greensand and Gault Lower Greensand Weald Clay Hastings Sands	Chalk at top Sandstones and Clays below
	Jurassic	Purbeck Beds Portland Beds Kimmeridge Clay Corallian Beds Oxford Clay and Kellaways Rock Cornbrash Forest Marble Great Oolite with Stonesfield Slate Inferior Oolite Lias—Upper, Middle, and Lower	Shales, Sandstones and Oolitic Limestones
	Triassic	Rhaetic Keuper Marls Keuper Sandstone Upper Bunter Sandstone Bunter Pebble Beds Lower Bunter Sandstone	Red Sandstones and Marls, Gypsum and Salt
PRIMARY	Permian	Magnesian Limestone and Sandstone Marl Slate Lower Permian Sandstone	Red Sandstones and Magnesian Limestone
	Carboniferous	Coal Measures Millstone Grit Mountain Limestone Basal Carboniferous Rocks	Sandstones, Shales and Coals at top Sandstones in middle Limestone and Shales below
	Devonian	Upper Middle } Devonian and Old Red Sand- Lower } stone	Red Sandstones, Shales, Slates and Lime- stones
	Silurian	Ludlow Beds Wenlock Beds Llandovery Beds	Sandstones, Shales and Thin Limestones
	Ordovician	Caradoc Beds Llandeilo Beds Arenig Beds	Shales, Slates, Sandstones and Thin Limestones
	Cambrian	Tremadoc Slates Lingula Flags Menevian Beds Harlech Grits and Llanberis Slates	Slates and Sandstones
	Pre-Cambrian	No definite classification yet made	Sandstones, Slates and Volcanic Rocks

worn surface of the older beds, the edges of which will abut against the oldest of the new set of flatly deposited beds, which latter may in turn undergo disturbance and renewal of their upper portions.

Again, after the formation of the beds many changes may occur in them. They may become hardened, pebble-beds being changed into conglomerates, sands into sand-stones, muds and clays into mudstones and shales, soft deposits of lime into limestone, and loose volcanic ashes into exceedingly hard rocks. They may also become cracked, and the cracks are often very regular, running in two directions at right angles one to the other. Such cracks are known as *joints*, and the joints are very important in affecting the physical geography of a district. Then, as the result of great pressure applied sideways, the rocks may be so changed that they can be split into thin slabs, which usually, though not necessarily, split along planes standing at high angles to the horizontal. Rocks affected in this way are known as *slates*.

If we could flatten out all the beds of England, and arrange them one over the other and bore a shaft through them, we should see them on the sides of the shaft, the newest appearing at the top and the oldest at the bottom, much as in the table annexed. Such a shaft would have a depth of between 50,000 and 100,000 feet. The strata beds are divided into three great groups called Primary or Palaeozoic, Secondary or Mesozoic, and Tertiary or Cainozoic, and the lowest Primary rocks are the oldest rocks of Britain, and form as it were the foundation stones on which the other rocks rest. These are usually termed

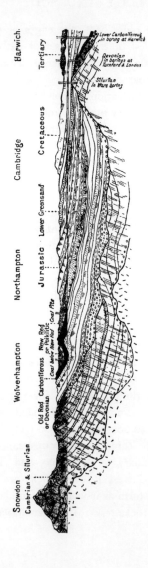

Diagram Section from Snowdon to Harwich, about 200 miles

Snowdon
Cambrian & Silurian

Wolverhampton

Old Red Carboniferous New Red
or Devonian   or Poikilitic
Coal below New Red "Coal Pits"

Northampton

Jurassic   Lower Greensand

Cambridge

Cretaceous

Harwich.

Tertiary

Lower Carboniferous
in boring at Harwich

Devonian
in borings at
Turnford & London

Silurian
in Ware boring

the Pre-cambrian rocks. The three great groups are divided into minor divisions known as systems. The names of these systems are arranged in order in the table. On the right hand side, the general characters of the rocks of each system are stated.

Let us now look at the nature of the rocks in our own county.

We must not expect to find rocks of all ages in one small portion of an island. The map at the end of the book shows by its colouring what kinds of rock are to be found at the surface in Staffordshire, and a key to the colours is given. On comparing our map with this we see that the igneous rocks are represented to the south-west of the Potteries by a " dyke " or solidified mass of dolerite, which was forced up when in a molten condition. Other isolated patches occur in the Black Country. In the south of the county various patches of old Wenlock shales and limestones show themselves, and are surrounded by rocks of the Carboniferous period, except where, owing to a crack or " fault,' the beds have been suddenly altered in level and the newer marls or sandstones of the Permian system have been brought into contact with the older Wenlock beds.

Passing northwards over the outcrop of the coal measures, we find that the Carboniferous strata lie underneath the sandstone and pebbly rocks of the Bunter period, which are particularly coarse near Wootton. From Cannock Chase the land dips down on three sides to the broad vale of Trent, which owes its gently undulating nature to the softness of the red Keuper marl, which is

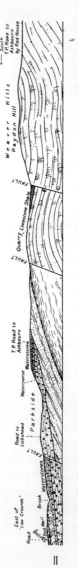

Geological Section across part of North Staffordshire from Smalley Farm to the Weaver Hills

readily washed into the rivers except where layers of harder marl and shale occur.

A little older than this red marl is the Keuper sandstone (f5 on the map) which underlies the marl and outcrops in a few places. These sandstone beds are often called the "Waterstones," for, since they are porous, and

Peakstone Rock

the beds above and below allow water to pass through with difficulty, they are saturated with water, which is pure though somewhat "hard," and is obtained in large quantity by sinking wells through the overlying strata. They also provide good building materials, but the economic importance of the rocks will be fully discussed later.

The cement with which the grains are bound together varies in different parts of the outcrop. Where it is relatively insoluble the rock is able to resist the agents of denudation for a long time : such portions stand out as hills or isolated rocks. The Peakstone Rock, near Bradley, resembling a sea stack left by the waves, is a remarkable example ; its cement is of barium sulphate, an exceedingly insoluble substance. The Himlack Stone, near Nottingham, is a parallel example.

In the north of Staffordshire the older Carboniferous system again makes its appearance. The limestones of this system rise up into the Weaver Hills, and continue thence into the Peak District of Derbyshire and the Pennine Chain.

These various kinds of rocks are usually covered up by the soil and vegetation, and it is only when this upper layer is removed for some purpose, such as the digging of foundations, quarrying, and railway cuttings, that the rocks are exposed for our study.

The character of the soil in some parts of the county cannot, however, be accounted for in this way. A certain kind of soil is found in patches, large and small. It is sometimes clayey, frequently full of pebbles and boulders, some of which are scratched, and very often these boulders do not resemble the underlying rocks, and have evidently been brought from a distance. The well-supported hypothesis that Northern Europe, including all England except the southern counties, was once covered with a thick glacier, accounts for this and other phenomena. The ice carried with it much dirt and many stones which scratched the rocks over which they passed and were themselves scratched. When the ice melted, this boulder clay or "glacial drift" was left behind in a layer of varying thickness and character.

We must not expect to find that the layers of which the earth's crust is composed are everywhere horizontal Movement and folding have taken place everywhere, and at present we are on a portion of a huge anticline, as our section shows.

The rocks dip towards the east. The upper portion of the anticline has been washed away or denuded, and now, as we travel across England from east to west, we pass in general across outcrops of older and still older rocks. Such a series of movements, however, could not be entirely regular, and numerous local folds, cracks, and tiltings of the strata are of course often met with. The strata, too, do not always possess the same character in all districts. What we have learnt of their formation clearly indicates that their thickness and nature must differ with the place and time of deposit. A great deal is taught as to their origin by the fossils, or remains of living things, which are to be found in them. On the next page is a photograph of a few plant-remains found in Staffordshire rocks.

From the paucity of the organic remains of the Triassic period geologists suppose that the regions where these deposits occur were of the same nature as the desert tracts of Central Asia. Layers of ripple-marked hard red and green marl are frequently found, while the chemical deposits indicate that an inland lake, possibly like the Dead Sea, occupied much of Staffordshire at this remote epoch.

Coal Fossils in the Denstone College Museum

1—7 *Alethopteris sp.* 8 *Lepidodendron sp.* 9 *Mariopteris muricata.*
10 *Stigmaria ficoides.*

7. Natural History.

The British Isles have not always been islands.   All round our coasts, at various places, we find the remains of submerged forests.   From various borings that have been made we have proof that there are ancient river channels at a great depth below the present sea-level, and fresh-water shells and bones of land animals have been dredged up in the middle of the North Sea.   The latter is, indeed, very shallow; so much so that if it were possible to place St Paul's Cathedral in it, the dome would be well above the water.   At one time, then, as we have seen, our land formed part of the Continent and doubtless had a similar fauna and flora.   But this is by no means the only change of level that has taken place, for an examination of some of our mountain slopes reveals an even greater movement in the opposite direction and we find sea-beaches and shells at the height of several hundred feet.   This great sub-mersion seems to have taken place during the latter part of the Glacial Epoch, and the elevation and union with the Continent that subsequently occurred were probably not—geologically speaking—of very long duration.   The submersion, like the ice-cap of the Glacial Period, must have destroyed our fauna and flora almost entirely, and the re-stocking with species would have to take place gradually in a north-westerly direction from the unflooded part of the Continent.   But subsidence again took place and the North Sea and the Channel intervened before all the Continental species had made their way into England. England is therefore poorer in its fauna and flora than the

Continent, and because it was separated still earlier Ireland is poorer than England.

If we consider the position and conditions of Staffordshire we shall not expect it to be peculiarly rich either in its zoology or its botany. It is, to begin with, an inland county, and shore-loving plants and sea-birds are absent from its lists or only of abnormal occurrence. Moorland it has, it is true, but moorland species, either of plant or animal, are not specially numerous. It cannot, like other inland counties such as Huntingdonshire and Cambridgeshire, boast of a fen-land, so rich in this respect; nor, like the southern counties of England, can it expect to receive numerous Continental stragglers. But it has one or two special points of interest. Firstly, it can show an extensive limestone district, and in this we always get a peculiar and varied flora. Again, it is in part on the line of an ancient migration track, running north-east and south-west, between the Humber and Wash on the one hand and the Bristol Channel on the other, and in consequence it is visited by great numbers of migratory birds. And lastly it is of interest, as we shall see, as being a sort of border county between northern and southern species, able to claim both the grouse and the nightingale.

Of the Staffordshire mammals there is not much to be said. Though there is still a good deal of wild country left, it no longer shelters, as formerly, the wild cat or the pine marten, both being now extinct. The last polecat, too, is stated to have been killed in 1884. The badger is not numerous, but is said to be increasing owing to protection. The otter still occurs in most rivers, but is not numerous and is much persecuted. All the British insectivores are found, the hedgehog, mole, and shrew abundantly, though the pygmy shrew and water shrew are more local. Truly wild fallow deer still exist in limited numbers in Cannock Chase, and in many parks they are, of course, kept as semi-domesticated animals, as are, in some cases, red deer. Until recent years the most interesting feature in the county was the Chartley

Wild Goats in Bagot's Park

herd of wild white cattle ; actual descendants of the wild cattle emparked in the reign of Henry III. Unfortunately tuberculosis made its appearance about 1900, and, except for a few head removed to Woburn Abbey in 1905, the herd became extinct.

Dr McAldowie has well pointed out that the county is a sort of border land where species of the north and of the south touch. Characteristic and not unusual features are the dipper and the ring-ouzel, and in Cannock Chase black-game and grouse, thanks to protection, are now abundant, though at one time nearly extinct. In the moorland, too, the twite nests. All these are characteristic northern forms. But we find also the nuthatch, the nightingale, and the reed-warbler, which are distinctly southrons. Of 216 birds recorded from the county, 103 nest, among them the tufted duck, an uncommon species, which breeds regularly. The bearded tit, one of the most beautiful and uncommon of British birds, now practically confined to the Norfolk Broads, has been recorded from Aqualate, and the raven was observed as lately as 1894. The merlin certainly breeds in the county and the hobby probably does so. The hawfinch is increasing, and the little-owl—no doubt descendant of the birds released from the Lilford aviaries in Northamptonshire—has apparently now established itself in the county. There are four heronries, at Aqualate, Patshull and Bagot's Parks, and near Cheadle.

Staffordshire is rather poor in butterflies, and only 42, or little more than half the total of British species, have been recognised, and of these none but the very commonest species are abundant. Of the reptiles the lizard is not uncommon, especially in Cannock Chase and in the north, and the blindworm, grass snake, and viper are also no rarities. Since Izaak Walton's day the streams have much altered, in no way, it may safely be said, for the better, and pollution from the factories and various works has done much harm, though of late this has been checked, with the result that fish are on the increase. Still, besides the ordinary coarse fish, there are trout and grayling in many rivers, and the salmon, it is said, still runs up the Dove.

Turning to botany we find the county somewhat richer. The land-surface is tolerably diversified. The high land in the north attains an elevation of some 1700 feet, and here and at Cannock Chase, and elsewhere, we get the moorland type of vegetation, such as ling, heath, the bilberry, crowberry, and whortleberry. The limestone region is rich in plants delighting in this soil, and Dove Dale is a favourite resort of botanists. The two violas, *Viola hirta* and *V. lutea*, the rock-rose (*Helianthemum*), the bitter-cress (*Cardamine impatiens*), Jacob's ladder (*Polemonium cœruleum*) and the sandwort (*Arenaria tenuifolia*), to mention only a few, are characteristic of this soil.

In several parts of the county brine springs exist, and in these neighbourhoods it is interesting to note the occurrence of maritime plants, which are possibly survivors of a pre-existent maritime flora. Among these are the sea-milkwort (*Glaux maritima*), sea-starwort (*Aster tripolium*), wild celery (*Apium graveolens*), the golden dock (*Rumex maritimus*) and the stork's bill (*Erodium maritimum*).

Although the increase of mining and other industries has greatly altered natural conditions, woodland still forms one-twentieth of the whole county. To the south-east is what is left of the once great Forest of Needwood, and here are still to be found interesting plants such as the mezereon (*Daphne mezereum*), the needle-furze (*Genista anglica*), and the small-leaved lime-tree (*Tilia parvifolia*), while the woodlands of the north especially have many rare plants. The bog lands have been a good deal reduced of late, but still yield the rarer water-plants, such as the bladderwort (*Utricularia*), grass of Parnassus, water-soldier (*Stratiotes*) and the long-leaved sundew (*Drosera longifolia*) besides various rare mosses. Indeed, the moss flora of Staffordshire, as far as is yet known, is more numerous than that of any of the surrounding counties.

8. Climate.

The climate of a country is the average of its weather. England is remarkable for the great tendency to variation exhibited in this respect at almost every season of the year. This is due in great measure to the fact that the British Isles lie in the path of storms of a cyclonic, or circular, nature which are constantly travelling in an easterly direction from the Atlantic. In these the mass of air whirls round a centre where the atmospheric pressure is low. In all cyclones in the northern hemisphere the twist is always against the hands of the clock. More than half the cyclones or "depressions" passing over the British Isles travel in such a way that their centres are north of Staffordshire.

Cyclones vary in shape and in speed. If the "gradient is steep," that is, if the barometer shows a large difference in pressure between the outside and the centre of the storm, then the storm area is usually somewhat restricted, but the speed at which it moves is great, and the velocity of the whirling wind may be very high. Staffordshire is rarely visited by storms of great velocity, and the effects of high winds are softened by friction with the varied surface. In this country "depressions" are generally shallow and wide-spread, and may take several days to pass over a given spot.

Let us follow the passage of such a storm over, say, Stafford. As it approaches, the barometer gradually falls, the wind blowing from the south-east. Before long the sky clouds over, rain begins to fall, and the wind veers round towards the south. As the centre draws near the barometer ceases to fall, but the rain continues, even when the centre has passed and the barometer rises. This rise is a sure sign that the depression has half gone, and soon the rain ceases, the wind now blowing from the west. At the end of the storm the sky is clear, while a cold wind blows from the north-west until normal conditions prevail, or another storm approaches.

Anticyclones are of the converse type, and are accompanied by cloudless skies, and winds that are frequently from the east.

Since cyclones and anticyclones pass over the country with great frequency, it is clear that the weather must be continually changing everywhere in our islands. Moreover, the prevailing winds come from the Atlantic, whose upper layers near our shores consist of the warm waters of the Gulf Stream or North Atlantic Drift. Hence they are warm and moist, and our climate has no great variations of temperature. The fact that the direction from which

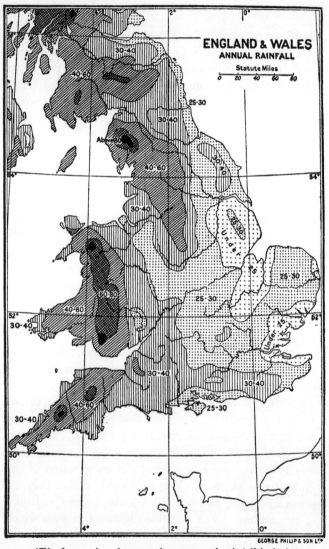

ENGLAND & WALES
ANNUAL RAINFALL
Statute Miles

(*The figures give the approximate annual rainfall in inches*)

the winds in this country most frequently blow is the south-west is shown by the bent growth of trees which are in exposed situations. So definite and regular is their north-easterly slope that the points of the compass may be roughly obtained by their aid.

The amount of rain a district or a country receives depends upon its nearness to the sea and to its height above sea-level. If we examine the rainfall map of England and Wales we see that the hilly regions of Cornwall and Devon, Wales and Cumberland receive most rain. The majority of the cyclonic disturbances which traverse these islands come from the south and west, so that before they reach the Midlands they have to pass the high lands of Wales, and on these the greatest precipitation takes place. On the Snowdonian mountains over 80 inches of water (including snow) fall per annum, while only 40 miles to the east lies the 30-inch isohyetal (the line joining places which have an average rainfall of 30 inches). The southern parts of Staffordshire, screened by Wales, receive less than 30 inches of rain per annum, the average at Wolverhampton (Wrottesley) for the years 1875–1909 being only 26·39 inches. The northern parts are influenced by the nearness of the southern extremity of the Pennine Chain and, as we should expect, have a larger rainfall. The 35-year average just to the east of Market Drayton is 29·41 inches, while at Cheadle it is 32·65 inches. The Moorlands have an average of about 40 inches.

Besides being wetter, the Moorlands, owing to their height, are colder and more wind-swept than the rest of the county, and snow is more frequent, deeper, and lies longer there than elsewhere, though Cannock Chase is very bleak in winter.

9. People—Races. Dialect. Population.

The Romans have left us descriptions of the people who lived in Britain when they conquered this country, but of the races who occupied the land before the Britons we have no written record. For our knowledge of them we are therefore restricted to studying the remains they have left behind them.

Geologists and archaeologists divide this pre-historic period into four ages. The oldest is the Palaeolithic or Old Stone Age, before the Straits of Dover had been formed and when the Trent was probably flowing into the Rhine. The men of this period used weapons and implements made of stone, very roughly worked. These people often lived in caves, and in close proximity to the flint implements and evidences of fire, the bones of such animals as the cave hyaena, cave bear, mammoth, and woolly rhinoceros are found. The remains found indicate that the climate of that day must have been very different from ours, so that the time which has elapsed since the Old Stone Age must be many thousands of years. This is borne out by the thickness of the stalagmitic layers which sometimes cover these bones and implements, and which form at an extremely slow rate. As far as is known at present Palaeolithic man has left no trace of his occupation in Staffordshire.

What gap, if any, intervened before the coming of the Neolithic or New Stone Age is a disputed question, but there is no doubt that the people of this latter period were of much more advanced culture than their predecessors. Their stone implements were better worked and were frequently ground and polished. The lake-dwellings, or houses built on piles driven into the bottoms of lakes, as in Switzerland and Scotland, belong to this period. Probably, too, the men of the New Stone Age kept cattle, sheep, and other domestic animals and cultivated the soil.

In Staffordshire there are some pits, at Seisdon, which may have been dwellings of this period, but it must be remembered that the district could have supported only a very scanty population before the draining of the marshy river valleys. The forests, however, would form welcome shelter for the people.

Neolithic man was succeeded by Celtic races who had learnt to fashion implements of copper or bronze. Their superior civilisation probably enabled them to drive out their stone-using predecessors. When the men of the Bronze Age buried their dead chiefs they covered them with great mounds of earth, known as barrows, which were often tribal burial places. Their skeletons show that they were taller than their predecessors and had round skulls. Many barrows are to be found in the Moorlands. In one of them, Mouse Low, a bronze drinking-cup was found, containing a flint arrow-head. In Thor's Cave, Wetton, articles of stone, bronze, and iron have been found in the same deposit. From these finds we may argue that in Staffordshire Neolithic man was only gradually superseded by the newer race using bronze, and that, in turn, the Bronze Age passed gradually into the Iron Age.

The Iron Age men were also of Celtic origin and lived here just before and during the Roman occupation. These Brythons or Britons became known in the eastern midlands as the Cornavii and as the Ordovices in the west. Julius Caesar tells us something about these people. In the south of England they possessed a fair civilisation, for they owned cattle, cultivated corn, used chariots in war which were armed with scythes, and even had a few war-ships. Coins were used by some of the tribes. They exported tin and used much iron. In the midlands the influence of the oversea trade with the adventurous Phoenicians, Greeks, and Carthaginians was hardly felt, and no corn was grown. The people wore skins and lived on milk and meat.

The Romans intermarried very little with the Britons, so probably did not greatly alter the native character. On the other hand, the Britons were not allowed to fight during the Roman occupation, and as a consequence lost their warlike nature. The Angles, Jutes, and Saxons had little difficulty in overcoming the Britons when they invaded Britain after the withdrawal of the Romans. The advance of the Angles was along the Trent and its main tributaries. The Britons living in the moorlands were thus cut off from their fellows and were unable to flee into the mountainous regions of the west.

By the Treaty of Wedmore between Alfred and the invading Danes in 878, the Danes, as we shall read later in chapter 12, were given the tract of land which lay to the north and east of Watling Street. Thus two-thirds of Staffordshire was within the Danish sphere of influence or Danelagh, but they never permanently settled in this county, and there are no place-names of Danish origin.

The Staffordshire dialect is a development of the Mercian or Midland form of Old English. There are distinct differences between it and the dialect of the bordering counties of Leicestershire and Warwickshire. These differences are probably due to the fact that the Angles, who came up the Trent, were in the majority in this region, while farther south the Saxon influence begins to be felt. In Staffordshire we notice that the letter *i* is pronounced *oi*, thus *life* becomes *loif*; *an* is pronounced *on*; *man*, *bank* become *mon*, *bonk*; *a* is pronounced as *ey* in *prey*, thus *father* becomes *feyther*.

In the preface to the *English Dialect Dictionary* George Eliot states that her intention when writing *Adam Bede* was to describe "the talk of North Staffordshire and the neighbouring part of Derbyshire." Ellastone was the central scene of the novel; one frequently hears there and in Denstone and other neighbouring villages such expressions, mentioned in the book, as "thee munna" (thou must not), "thee wouldstna," "thysen," "I shanna," "donna thee," "if he warna," and so on.

The place-names of Staffordshire show many traces of their Anglo-Saxon origin. Many names end in *ley*, meaning pasture or field, e.g. Bradley, Ashley, Dudley. *Ton* means town, as Forton and Wolstanton. There are many *halls*, e.g. Tettenhall, Pelsall. The *lows* (burial mounds of chieftains) seem to have retained their names through Anglo-Saxon times. *Ray* is possibly a corruption of the Latin *regis*, e.g. Blore Ray. *Ey* means isle,

"Adam Bede's" House

Chebsey = Ceobba's Isle. *Was* = a marsh or swamp, e.g. Alrewas = Alder swamp. *Ville* is often a late corruption of field, e.g. *Enville*. *Wich* means village, e.g. Hammerwich.

The present population of Staffordshire is of mixed Norman and Anglo-Saxon descent, the Anglian strain predominating, with a slight admixture of Danish and

Celtic blood. Since the Norman Conquest the infusion of foreign blood into Staffordshire has been slight, and insufficient to change the character of the population. Living on Biddulph Moor are shy people with red-gold hair, the descendants possibly of a party of Saracens who came with returning Crusaders.

At Abbots Bromley an interesting survival is seen in the folk-dances held every September.

Horn Dancers, Abbots Bromley

The enormous industrial development of Staffordshire during the last hundred years has caused a great increase of population in the Pottery and Black Country districts, but the country villages have not increased to anything like the same extent, even diminishing in many cases. The county has the fourth largest population of all the English counties, with a density of 1164 persons to the square mile ; it has increased with remarkable regularity; the population of the Moorland region is scanty, but the recent opening up of the beautiful Manifold valley by a light railway will probably result in increased quarrying of the limestone and consequent influx of workmen to the district. The Moorlanders are frugal people. Formerly they used to eat oats of their own growing, making excellent porridge (known as "lumpy-tums") by stirring oatmeal into boiling milk. Flat sour oatcakes baked on a stone ("bak-stone") are still made in the district.

10. Agriculture.

The area of the county of Stafford is 741,320 acres, as last revised by the Ordnance Survey : this is made up of 734,920 acres of land, and 6400 acres of water surface. Mountain and heath land used for grazing amounted to 8288 acres in 1913, and 38,860 acres were covered by woods and plantations.

The reader is referred to the diagrams at the end of the book to obtain a better idea of the proportions which the areas occupied by various crops bear one to another and to the whole. On looking at these diagrams we are at once struck with the large proportion of grazing land, almost 74 per cent. of the acreage under crops and grass, which is in turn 80 per cent. of the land surface.

On the whole, the area of agricultural land in the country is diminishing : this is necessarily the case as towns grow larger, and ground is taken for railways, reservoirs, and other non-agricultural purposes. For many years there has been a steady increase in the amount of permanent pasture, and a decrease in the area of land under the plough all over the United Kingdom. In 1912, however, there was an increase of 36,000 acres in the area of arable land, and Staffordshire had a fair share in this improvement.

Grazing is the chief agricultural feature of the county. The cattle are mainly Shorthorn ; the sheep of the Cheviot, Hampshire, or Shropshire breeds. In 1913 there were nearly three acres of pasture to every cow, and one sheep to every 2·32 acres ; the figures show that there were 2000 fewer cattle and 14,000 fewer sheep than the year before. Much condensed milk is manufactured in the county. There are a few cheese factories in Staffordshire, but a great deal of the milk is sent up to London and other large towns. In 1913 there were over 38,845 pigs in the county, and several large bacon-curing factories are established in the Black Country.

The orchard area is not great, but the damsons are famous, and large quantities are sent every year to jam manufacturers. Careful cultivation is, however, seldom seen, and most of the orchards are grass-covered and serve as pasture.

While in every division of the county there is more pasture than plough land, the variation between the proportions of the two is very marked. In the Leek district

there are 20 acres of pasture to one of plough land, in
Uttoxeter 16 to one, and in Cheadle 11 to one. These
comprise the north-east of Staffordshire. West and south
of these the ratio gradually diminishes till in the Kingswinford and Wordsley, Wolverhampton, and the Lichfield and Brownhills districts it is less than two to one,
and in the Tamworth, Wednesbury, and Penkridge areas
less than three to one. Rather less than half the arable
land is devoted to corn crops. Owing to a suitable soil
and a good market at Walsall for the straw (which is used
for stuffing saddles), oats were the principal corn crop,
being grown on 29,953 acres. In this county nearly 3000
fewer acres of wheat were grown in 1913 than in 1912.
The average yield over the years 1902–1911 was 31·87
bushels per acre, while that of Northumberland was 35·18,
of Kent 35·09 and Cambridge 32·81. Staffordshire is
slightly above the average of England and Wales, which
is 31·71 bushels per acre. Barley occupied 15,844 acres
in 1913, mainly near Burton.

Of smaller crops we may notice that, in 1913, only
8¼ acres of strawberries and 11¾ acres of raspberries were
under cultivation in Staffordshire. This is curious, for in
the neighbouring county of Salop, with a slightly smaller
area, there were 29¾ and 19½ acres respectively, and in
Derbyshire, which is smaller, 117 and 91¾ acres respectively were devoted to these fruits.

In June 1913 the number of separate agricultural
holdings was 11,869, and of these 22·52 per cent.
were between one and five acres, 72·08 per cent. above
one and not exceeding 50 acres, and 25·98 per cent.

above 50 and not exceeding 300 acres, and 1·93 per cent.
were above 300 acres. Only eight counties of England
have a smaller percentage of farms of over 300 acres.
Only 8·48 per cent. of the acreage under grass and crops
were occupied by their owners, this being less than the
proportion for England generally.

11. Industries and Manufactures.

Probably no other county has so many industries as
Staffordshire. The iron trade has assumed vast proportions in the now unlovely district appropriately called the
"Black Country." This includes the towns of Smethwick, West Bromwich, Dudley, Oldbury, Sedgley,
Tipton, Bilston, Wednesbury, Wolverhampton, and
Walsall, so that it does not lie wholly in Staffordshire.
Beneath the surface are thick seams of coal, while iron
ore and clay exist in plenty. The district has been
famous on account of its iron-works for about four
hundred years, though until the middle of the eighteenth
century most of the fuel for smelting was charcoal, which
was obtained from the neighbouring forests. Coal smelting had been introduced much earlier by Dud Dudley,
who set up several furnaces in 1619, but the value of the
new method seems not to have been at once appreciated.
The occurrence of coal and iron-ore in such close
proximity has resulted in the growth of an enormous
industry.

The iron-ore, much of which is nowadays imported from Spain and elsewhere, is frequently roasted to expel water, carbon dioxide, and sulphur, and to make the fragments more porous so that the gases in the smelting furnace can act more freely.   The roasting or calcination is effected either in the open air or in specially constructed

Staffordshire Blast Furnaces

kilns, fuel being mixed with the ore if necessary. Certain ores, however, such as the black band iron-stone, contain so much carbonaceous matter that no additional fuel is required.

The calcined ore is transferred to the blast furnace. This is a large erection like a round tower, constructed at the present time of iron lined with fire-brick and varying in height from 70 to 100 feet. The furnace, when started, is dried by the burning of wood-charcoal inside it before being filled with coal, the heat from which is increased by means of *tuyères*, or twyers, delivering a blast of air at the bottom of the furnace. Charges of ore mixed with limestone or other flux added to remove impurities from the iron, alternating with fuel, are introduced through the top as required. The reduced iron takes up carbon and collects at the bottom, whence it is run off in the desired quantities into sand moulds, or taken to casting machines. The hot gases formed during the process used to be allowed to escape into the air from the top of the furnaces, but are now used to heat the blast, calcine fresh ore, or drive engines for producing the blast.

For the production of steel and varieties of iron, the crude metal, now in short bars called *pigs*, is subjected to purifying processes, and the percentage of carbon it is allowed to retain is regulated very carefully.

Each town of the Black Country is the seat of manufacture of some special kind of hardware. Thus locks are made at Wolverhampton and Willenhall, keys at Wednesfield, harness fittings and saddlery at Walsall and Bloxwich, gas fittings at West Bromwich. Of heavier goods, iron-work for railway carriages is manufactured at Wednesbury ; anchors, cables, and pumping machinery at Rowley Regis and Tipton ; tinplate and japanned ware at Bilston and Wolverhampton ; and chains and nails at Cradley. Tin toys, pins, spectacle frames, boilers, bicycles, mouse-traps, motor-cars, gunlocks, hammers, anvils, saws, spades, and a host of other articles in metal work are turned out of one place or other in the Black Country.

In the Black Country, too, are works for the manufacture of soda, sulphuric acid, and other chemicals.

The mineral wealth of Staffordshire is very great. In the year 1911 no less than 14,049,512 tons of coal were raised by the 56,270 persons employed in connection with the mines of the county. The coal-fields are five in number. In the extreme north-east there is the Goldsitch with an area of only 90 acres. The Shaffalon field is a little one, of small value, between Cheadle and the Potteries. The Cheadle coal-field is in the Churnet valley. It is said to contain nearly 100,000,000 tons of unworked coal. A triangle having Biddulph, Longton, and Madeley as its points may be considered roughly to be the boundary of the North Staffordshire or Pottery coal-field. The coal-measures here are about 5000 feet thick, though probably not more than 1200 feet consist of coal-seams or iron-stone beds. There is an enormous quantity of coal in this field available for future use. The South Staffordshire coal-field is in two portions, the Cannock and the Dudley. In the latter is the famous Thick or Ten-yard Coal which comprises from 10 to 14 well-marked seams, forming a mass of coal unequalled elsewhere in the British Isles. Recently there has been a great extension of the South Staffordshire coal-fields, e.g. at Baggeridge and near Hollybank (Essington); in each case coal has been found at the other side of the great fault.

Beds of iron-ore and fire-clay are frequently found associated with the coal-seams, and are worked to the great benefit of the industries of the county. Stafford-shire ranks second among the counties for the production of iron-ore.

The Triassic rocks, in the Needwood Forest district, contain valuable deposits of gypsum, or calcium sulphate, some of it in the compact form known as alabaster. Quarries have been worked here for several hundred years. It is said that the monks of Burton Abbey discovered that the local waters produced good beer. Certainly Burton has long been famous for the excellence of its brewing, upon which the inhabitants depend for their livelihood. It is said that the large proportion of calcium sulphate in the water precipitates the solids and gives a clear ale.

There are copper-smelting works at Smethwick in the Black Country, and at Froghall in the Churnet valley; while at Oakamoor, between Alton and Froghall, the beauty of the scenery is spoilt by the smoky chimneys of the large copper wire and tube mills. The first Atlantic cable was made here. In the Moorlands one comes across deserted copper and lead mines. Buxton Crescent is said to have been built with the profits made by a Duke of Devonshire from the copper mines at Ecton (long since disused) in the lovely Manifold valley.

At Caldon in the Weavers there are limestone quarries. The Triassic sandstones which outcrop at Alton and Hollington provide good building material, and mill and scythe-stones are also made from them.

The igneous rocks of the south are used for road-mending, and the sand at Bilston for making moulds in which iron is cast.

Large glass-works exist at Tutbury, West Bromwich, Smethwick, and Wordsley. The manufacture of boots and shoes gives employment to great numbers at Stafford and Stone. Huguenot refugees settled in Leek and its neighbourhood, bringing with them the knowledge of silk-working. The industry has prospered exceedingly, and now there are several mills of considerable size devoted to the weaving and dyeing of silk goods. Certain firms have their own silk farms in China, and some of the raw material comes from Japan and Bengal. Recently the manufacture of artificial silk has been introduced. Wood pulp is imported in the form of slabs like thick cardboard, and is worked up at Coventry into yarns which are then utilised by the Leek mills.

Braid is made at Leek: there are cotton mills at Rocester and Mayfield (near Ashbourne). Tape is made at Tean, near Cheadle, and textile small ware at Cheadle, Fazeley, and Tamworth, while Fazeley has also a bleaching mill.

Salt works are carried on at Baswich on the north-western extremity of Cannock Chase: the brine is brought in pipes from Stafford Common. The blue bricks used for pavements are made at Cannock, from a clay containing a high percentage of iron oxide.

The manufacture of pottery, carried on to an enormous extent in the district in the north-west of the county known as the Potteries, was originally dependent

upon the local clay and coal. In the sixteenth century
rough pottery was made at Burslem. The clay used was
reddish and coarse in texture : after being decorated in
various colours, it was glazed and baked. The size and
quality of the large dishes, candlesticks, butter-pots, loving
cups with two or more handles, etc. were regulated by

A Thrower's Wheel, Burslem

an Act of Parliament passed in 1661 (13 and 14 Ch. II,
cap. 56). In 1626 a monopoly was granted to Thomas
Ram and Abraham Cullen for " the sole making of
the Stone Potte, Stone Jugge, and Stone Bottle within
our Dominions, for a terme of Fowerteene years." Lead
glaze was used at first, but was superseded by salt glaze
after its discovery in 1680. A little later two Dutchmen

named Elers settled at Bradwell and Dimsdale, and produced from a local vein of fine clay some excellent ware. They also worked out the process of manufacture of black pottery, using manganese as colouring matter. It is said that they preserved their secrets for a time by the employment of half-witted workers, but in 1710 they removed to London after the discovery of their processes by their followers, Astbury and Twyford. These two were the first to use clays imported from Devonshire and Dorsetshire. Nowadays very large quantities of potter's clay are brought by sea from these two counties and Cornwall to Liverpool, and thence by canal to the potteries.

By the end of the seventeenth century Burslem, " the mother of the Potteries," was celebrated for its Delft ware (an imitation of the ware made at Delft, in Holland, and soon to supersede it), Crouch and mottled pottery, and the misnamed " Elizabethan " ware.

The beautiful pottery of Josiah Wedgwood made Etruria famous. " Wedgwood " ware is produced at the same works, together with " Majolica," "Queen's " and " Rockingham" ware, and Parian statuary. A special maroon variety of Wedgwood or jasper ware, with white figures on a dark red ground, was manufactured at one time solely for America, but at a time of trade depression the demand ceased, and the ware is made no longer.

During the nineteenth century the growth of the industry has been rapid. At the present time many thousands of persons of both sexes are employed in the manufacture of all kinds of pottery, from the most

Wedgwood Potteries—Jasper-ware making

Working on the Portland Vase

delicate porcelain and china to heavy stone bottles and sanitary ware. Very great precautions are taken nowadays lest the workers should suffer from the dust which is produced during certain processes or from the lead compounds used in making glazes. Substitutes have been discovered for glazes containing lead, but it is said that if the rules of cleanliness are strictly observed lead poisoning need never occur.

The principal towns in the Potteries are Stoke-on-Trent, Hanley, Longton, and Burslem. Their corporations and the councils of Fenton and Tunstall were dissolved in 1910, and all are now united, forming one county borough which takes the name of Stoke-on-Trent.

In very few counties is there such a large variety of industries, nor are they usually so localised. In the Potteries especially it is possible to get away from the dirt and dreariness in a few minutes, and reach very pleasing scenery.

12. History.

We have already seen that Staffordshire was inhabited before Roman times, but our knowledge of the peoples who lived here and of their doings is derived from the study of their forts, their dwellings, their burial-places, and the implements they used. There are no written records of these people except the little we gather of the later races from Roman accounts, and they thus belong to the domain of the antiquary and geologist rather than the historian.

From written records we learn that the Romans, when they invaded England, subdued the Cornavii and Ordovices who lived in the Midlands, and established military stations or garrison towns at intervals. In Staffordshire these were few and far between. The largest was Letocetum, the modern Wall, near Lichfield. This was situated close to the crossing-point of two of the trunk roads which the Romans built across the country to facilitate the movements of their troops. One of these was the Watling Street, which ran across South Staffordshire in a nearly due east and west direction ; the other was the Ryknield Way, which entered the county near Birmingham and left it in the neighbourhood of Burton. Apart from these roads and the military stations which are now towns or villages, such as Rocester, Wall, and possibly Uttoxeter, the Roman occupation left little impression on our county. In the fifth century it came to an end, and the British, having lost the art of fighting, were easily overcome by the invading Angles, Jutes, and Saxons.

Although the Saxons came before the Angles, they were not as numerous, and Staffordshire was occupied chiefly by the latter. The Britons fled mainly into Wales and the Moorlands, but probably a remnant were able to remain among their conquerors.

The English tribes first occupied Staffordshire just before 600 A.D. and, as we have seen, probably made their way up the Trent and its tributaries. Christianity, first introduced into Britain in Roman times, was driven out to the north and west by the pagan English, and was re-introduced into Mercia from the north. Later, this Celtic christianity was met by that which spread upwards to Mercia from the south as a result of the missionary work of Augustine and his successors, and in time the combined English Church agreed to bring its customs into line with those brought by Augustine from the continent. St Chad was Bishop of Lichfield from 669–672; the diocese extending in his day from Lincolnshire to the Wye, and from the Humber to the Thames.

The English settlements in England gradually consolidated into about seven kingdoms, of which the Kingdom of Mercia, founded by Cridda about 584, comprised at one time Gloucestershire, Worcestershire, Shropshire, Cheshire, Staffordshire, Derbyshire, Leicestershire, Northamptonshire, Rutlandshire, Nottinghamshire, Lincolnshire, Warwickshire, Huntingdonshire, Bedfordshire, Buckinghamshire, Oxfordshire, and a part of Herefordshire. After much fighting amongst themselves these kingdoms were finally welded into one by Egbert, King of Wessex (the kingdom in the south-west) in 827.

The Danes now became a menace, and about fifty years after Egbert's conquest they had made themselves masters of the whole country. The English King retired into obscurity for a few years, but in 878, seven years after his accession, Alfred the Great collected his forces and defeated the Danes at Ethandune in Wiltshire. A Treaty was made at Wedmore in Somerset in 878, and the invaders were allowed to retain the country north and east of a line which roughly followed the Watling Street. Thus two-thirds of Staffordshire was for a time in the

Danes' country or Danelagh. Edward the Elder conducted operations against them and defeated them at Tettenhall in 910, and three years later Ethelfleda, his sister, rebuilt Tamworth and Stafford as defences against them. In 918 Ethelfleda died at Tamworth, which had been a favourite royal residence since the days of Offa. One hundred years later the county was ravaged by Edmund Ironside because he did not receive sufficient assistance from Eadric, Earl of Mercia.

In 1069 or 1070 the county was again harassed because it refused to acknowledge the authority of William the Conqueror. Domesday Book, that wonderful survey of England, completed in 1086, contains evidence that for some years Staffordshire was a very poor and sparsely populated district. For very many years its great forests provided much hunting for the Kings of England and their nobles. The Forest of Needwood and Cannock Chase were their especial playgrounds. King's Bromley, in Needwood, was a royal demesne under the Normans, and the Essex Bridge was built by Queen Elizabeth's favourite so that he could reach the hunting grounds on Cannock Chase more easily from Chartley. Needwood remained a forest until it was enclosed in 1801.

King Stephen in his quarrels with the Barons received much support from Henry de Ferrers of Tutbury and Sir Robert Marmion of Tamworth, but in the struggles of Henry the Third's reign the county sided against the King.

In John's reign the Papal Legate lodged at Burton Abbey and received there the protest of Archbishop

Tamworth Castle

Langton and the clergy against any invasion by him of the rights of the Church of England. Thus it may be said that the first step in the movement which resulted in Magna Carta was taken in Staffordshire. The Chronicle of Burton Abbey is of great historical value.

The Black Death caused terrible mortality in 1349 and again in 1361. During the Wars of the Roses Staffordshire supported the Lancastrians, probably because the Duchy held much land here. The Battle of Blore Heath (1459), in which the Lancastrians were defeated, was fought in the west of the county. Henry VII marched through Stafford, Lichfield, and Tamworth to the Battle of Bosworth Field, fought not far away in Leicestershire.

Many Staffordshire people, such as the Pagets, the Giffards, and the Wolseleys, suffered for their faith during the unhappy persecutions of Roman Catholics by Protestants and of Protestants by Romanists as each in turn secured the advantage under the Tudors. Mary Queen of Scots was imprisoned at various times at Tutbury, Chartley, and Tixall. At Tutbury one part of the castle ruins is called Mary's Tower, but the building she occupied stood in the tiltyard and nothing remains of it. The conspirators in the Gunpowder Plot were captured or killed at Holbeach House, Himley, whither they had fled.

There were no great battles in Staffordshire during the Civil War. The county as a whole was on the side of the Parliament. Lichfield Close, Eccleshall Castle, Tutbury Castle, and Dudley Castle—the last in

Worcestershire but enclosed in Staffordshire—were besieged by one party or the other during this period, much damage being done to Lichfield Cathedral. A skirmish took place at Hopton Heath, near Stafford, in 1643, when a Royalist force was defeated and its leader, the Earl of Northampton, was slain. At Uttoxeter the Duke of Hamilton and the remnant of his Scottish army surrendered to General Lambert in 1648. Boscobel and the Royal Oak are in Shropshire, just over the border, and part of the route followed by Charles II as he fled from the battlefield of Worcester lay through our county.

The town of Stafford supported Charles Stuart in 1715, but in 1745 "Prince Charlie" found a cold welcome when he marched across the Moorlands during his ineffectual advance, and again in his retreat from Derby. This retreat was by no means as orderly as the advance had been. The force was more scattered; plundering, forbidden during the advance, became common, and the town of Leek suffered considerably at the hands of the fleeing army.

13. Antiquities.

We have already seen that the period before the coming of the Romans has been divided into four "Ages," the Palaeolithic, the Neolithic, the Bronze Age, and the Iron Age. Numerous relics of prehistoric man of the Stone Age have been found in Staffordshire, but all are of Neolithic man. Most of them are instruments of flint when this was obtainable, or of hard stone, or, as in the Moorlands, of chert. They are of better workmanship and show a higher state of civilisation than do Palaeolithic remains.

Neolithic man, like his Celtic successors of the Bronze Age, heaped huge earthen mounds over the dead bodies of his chieftains, and often the interior of such a mound was the common burial place of his tribe.

These tumuli, lows, or barrows, are generally much longer than those of later date ; sometimes they are as much as 400 feet long by 80 feet broad. They are generally single-chambered and contain skeletons in a crouching position surrounded by the objects they prized in life. At first the dead were not burnt, but in the later Stone Age this practice was begun.

There are over fifty of these "lows" in Staffordshire, a large proportion of them occurring in the Moorlands ; for instance, the parish of Alstonfield contains no less than six, Farley has three, Ilam has four. Many of these, however, belong to the later Bronze Age, and are distinguished from the older barrows by their smaller size and rounder shape. The implements found in them are of stone and bronze, indicating a transition period, or even of bronze alone. While Neolithic man seems to have preferred the Moorlands, his successors were more evenly distributed over the county.

Iron does not seem to have been well known in the north of Staffordshire until historic times, for few iron articles are found in local deposits of Iron Age date. In a tumulus at Leek called Cock Low, demolished in 1907 to make room for buildings, an urn was found containing animal bones and small pieces of a child's skull, together with traces of woad which still retained its light blue colour.

In other parts of Britain, pit and cave dwellings of prehistoric man are sometimes to be seen, but in Stafford-

Thor's Cave

shire none have been discovered, unless we are to regard the pits at Seisdon as Neolithic homes.

In the limestone of the north-east there are numerous caves. Such caves were often inhabited by wild beasts, and man himself frequently lived in them. In Staffordshire we have evidence that at least one has been inhabited. Thor's Cavern, near Wetton, was undoubtedly occupied by the Romano-Britons as a permanent habitation. The discoveries in it, however, were all of the late Celtic period, nothing belonging to the age of stone or to the Saxon period being found.

Near Biddulph is an ancient burial place known as the " Bride Stones." It consists of a *kistvaen* or chest made of flat stones of local origin with accompanying monoliths, and resembles those in the circle of kistvaens near to Port Erin in the Isle of Man. The hill on which it is built appears to have been occupied by other burial places.

Remains of hill forts exist at Bunbury, just behind Alton Towers; Castle Old Fort, near Upper Stonnal; Bury Bank, near Stone; Bury Ring on Billington Hill, two miles south of Stafford ; Castle Ring, in Beaudesert Park on Cannock Chase ; Byrth Hill, near the Cheshire border ; and the village of Maer. All seven forts are of the same type and have similar characteristics, so may be assigned to the same date. " They appear as defended retreats for use in case of attack," writes one authority, " places where a whole community might resort, rather than as camps for stationary garrisons." They are all situated sufficiently high for a good view of the surrounding country to be obtained. The designers have made full use of the precipitous sides of hills as aids to fortification, digging fosses and raising ramparts wherever the natural features of the site did not afford sufficient protection. The approaches are in full view of the ramparts and the entrances are guarded by mounds. There is always a water supply within easy reach.

Inside Bury Bank is a central mound, which is said to have been built by Wulfere, King of Mercia, who very probably lived here, having taken possession of the old fort and adapted it to his purpose.

Passing from prehistoric times to the Roman occupation we notice first the roads built by the conquerors. The important trunk road known as the Watling Street entered Staffordshire at Fazeley Bridge, a few miles south of Tamworth, and passed close to Lichfield, where was the Roman station of Letocetum, now Wall. Watling Street turns due west here in order that it may avoid the forest uplands of Cannock Chase. After passing south of Cannock and close to the station of Pennocrucium it crosses Staffordshire by a series of straight stretches, each of which seems to have been aligned by means of sighting stations on hill tops, and leaves the county near Sheriff Hales. The Ryknield Way is lost nowadays among the streets of Birmingham, but becomes visible as it emerges towards the north-east. It crossed the Watling Street just east of Letocetum, and passing over the Trent near Burton, went straight on to Derby.

There is some evidence tending to show that another Roman road entered Staffordshire not far from Betley which went as far as Chesterton, a village which has been identified as the Mediolanum of the Antonine Itinerary.

A Ryknield Street is mentioned in the Charter of Abbey Hulton as one of the boundaries of the lands at Normacot Grange near Longton. This points to the existence of a Roman road connecting Chesterton and the eastern stations of Uttoxeter and Rocester. It is

Ilam Cross

quite possible that intersecting roads were made by the Romans to join up their stations directly, and in two cases in Staffordshire there are indications that such roads did actually exist. One of these led from Rocester to Derby, while Red Street, not far from Chesterton, is probably of Roman origin.

Most of these roads of certain or probable Roman origin are in use to-day throughout the greater part of their length. They are characterised by the directness of their course up hill and down dale, and were generally stone paved. It will be noticed that there is not one in the Moorlands region, so far as is known, though doubtless there were numerous tracks.

Roman Camps are still to be seen in Staffordshire. Those at Wall and Chesterton have already been mentioned. There is one near Green's Forge, in the south, and near Longdon Church, north-west of Wall, is another. On Barrow Hill, near Rocester, a camp overlooks the low-lying village. All these camps are rectangular, with rounded angles and a single rampart and fosse; and are not far from water. It is curious that Wall is the only one of them which is situated on a Roman road.

There are no remains of Saxon buildings known to exist in the county, but churchyard crosses of pre-Norman date are to be found at Ilam, Checkley, and other places. Saxon fonts, too, remain in several churches. These are of two kinds; in one the whole length of the shaft is rectangular, in the other a rectangular shaft springs from a round base.

14. Architecture—(a) Ecclesiastical.

We get so much accustomed to our churches, having grown up with them from childhood, that the majority of people, perhaps, give little thought to them, to their antiquity, and to the story they tell. Most of them, however, in their site if not in their actual fabric, are far older than the oldest houses, or even than the castles. They alone remain as relics of times long gone by, and in and around them have focussed the interest and history of the parish from the earliest times. While great families have sprung into existence, flourished perhaps for centuries, and then become extinct, the churches have remained, often untouched, the silent witnesses of the story of the parish from medieval days.

Before the coming of the Normans there were numerous churches in our land, but not many of these so-called Saxon buildings now remain, and none in an untouched condition. This is partly due, not merely to age, but to the fact that the builders of those days were more or less unskilled. They had an imperfect knowledge of construction and built walls of rough rubble unsupported by buttresses. The churches were thus necessarily small. The doorways and windows had small semi-circular or triangular arches, and there were square towers with what is termed " long-and-short " work at the quoins or corners. Staffordshire cannot show much in the way of Saxon or Pre-Norman masonry, though it is probable that certain portions of Tamworth Castle are of this period, and possibly some of St Chad's at Stafford.

Norman Doorway, West front, Tutbury Church

With the advent of the Normans an active period of building commenced, both of castles and churches. The leading characteristic of these is massiveness. The walls were of great thickness, the windows small. The vaulting was semi-circular, as were the heads of the doorways

St Editha's, Church Eaton, from the East

and windows; and massive towers, sometimes lofty in the case of cathedral churches, were frequent. This style, known to us as Norman, is termed Romanesque on the Continent, and was prevalent until the middle of the twelfth century. In spite of the very large number

of churches built before 1200, there is not much
Norman work in Staffordshire. Tutbury church, which
formed part of an old priory, Gnosall, and St Chad's in
Stafford, however, are more or less rich in this style, and
examples of it are to be found in about twenty other
churches scattered throughout the county. Of transitional
Norman, when that style was passing into the next, we
have a good instance in the tower at Church Eaton.

From about 1150 to 1200 the building became lighter,
the arches pointed, and there was gradually perfected the
science of vaulting, by which the weight is brought upon
piers and buttresses. This method of building—the
"Gothic"—originated from the endeavour to cover the
widest and loftiest areas with the greatest economy of
stone. The first English Gothic, called "Early English,"
from about 1180 to 1250, is characterised by slender
piers (commonly of marble), lofty pointed vaultings, and
long, narrow, lancet-headed windows. After 1250 the
windows became broader, divided up, and ornamented by
tracery, while in the vault the ribs were multiplied. The
greatest elegance in English Gothic was reached from
1260 to 1290. Good examples of Early English in
Staffordshire are the ruins of Croxden Abbey, the choir,
transepts, and chapter-house of Lichfield Cathedral, and
the chapel at Coppenhall.

After 1300 the structure of stone buildings began to
be much overlaid with ornament, the windows were
wider, and their tracery and the vault ribs of intricate
patterns, the pinnacles and spires loaded with crocket and
ornament. This style is known as the "Decorated," and it

Lichfield Cathedral—West Front

came to an end with the advent of the Black Death, which stopped all building for a time. Much good work of this period is to be seen in our county, notably the beautiful Lady Chapel and nave of Lichfield Cathedral, the greater part of Clifton Campville church, and the chancel at Checkley.

While evidences of transition are of common occurrence in the passing from one style to another, this is practically non-existent, for the reason just mentioned, between the Decorated and the "Perpendicular," which succeeded it. Unknown abroad, this style developed with curious uniformity all over England after 1360, and lasted with but little alteration up to 1540. It is characterised by the perpendicular (and horizontal) arrangement of the tracery and panels in windows and on walls, and also by the flattened arches and square moulding over them, by the elaborate vault-traceries, especially fan vaulting, and by the use of flat roofs and towers without spires. Of this style the finest instances in Staffordshire are the collegiate churches at Wolverhampton and Penkridge. After the Reformation ecclesiastical building nearly ceased, but two churches in our county, Blurton and Broughton, built in the time of Charles I, show how the country builders kept to the Gothic style.

As in other parts of England, so also in Staffordshire, the Tudor and Jacobean architecture is mainly of the domestic order. Flat-headed windows, level ceilings, and panelled rooms became the fashion, with classical ornaments in the later period. About this time the professional architect came into existence: hitherto building had been entirely in the hands of the builder and the craftsman.

The Nave, Penkridge Church

15. Architecture—(b) Military.

Mention has been made already of the forts which were built in British and Roman times on the tops of many hills in the county. Military architecture of Anglo-Saxon origin was spoken of in the first chapter, in which the building of the castles at Tamworth and Stafford in 913 is assigned to Ethelfleda, the warlike daughter of King Alfred, called the "Lady of the Mercians." These two castles doubtless did much to keep the Danes in check.

It is doubtful whether any of the present masonry at Tamworth is part of the original building, though the causeway and a portion of the mound may be Saxon. During Norman times the Marmions rebuilt the castle. Some of their rubble-work still remains in the tower and curtain-wall. Later, the de Frevilles effected many repairs with roughly-squared stone. At the close of the sixteenth and beginning of the seventeenth centuries the Ferrers family added a house built in the prevailing Tudor and Jacobean style. Much of this building is now incorporated in a museum.

Domesday Book mentions only the ruins of a castle at Stafford, so Ethelfleda's "burh," or "enclosed place," and a new building erected by William the Conqueror in 1070, had evidently been destroyed by 1086. William Rufus built a third castle, of which nothing remains, Sir William Brereton having destroyed all he could after capturing it from Lady Stafford, who held it for the King for a short time in 1643. The town itself was protected

Tutbury High Tower

by walls with four gates; of these only a fragment of the Eastgate remains, but the street names,—Greengate, Eastgate, Foregate and Gaolgate, and North and South Back Walls, remind us of what has been. The building now called Stafford Castle is only a sham castle, begun, but not finished, in the early years of the nineteenth century. The view from its towers shows how suitable the site was for military occupation.

Several castles have, at different times, occupied the site of the ruins at Tutbury. Duignan, in his *Notes on Staffordshire Place Names*, says that the word Tutbury is probably derived from the Anglo-Saxon verb *totian*, to project, hence to put one's head out, look around, spy, and *burh* an enclosed place or fort. The name Toot-hill is found to occur at various places, and their situations agree with the meaning " mount of observation." Therefore in all likelihood Tutbury was a military site in Saxon times. Henry de Ferrers built a castle here in the time of the Conqueror, but its destruction was included in the punishment of the family in 1174 for taking the part of Prince Henry against his father, Henry II. Later, the Ferrers family regained the site and rebuilt the castle, but it was destroyed in 1264, and again rebuilt in 1298 by Edmund, Earl of Lancaster. His son fled before Edward II, losing much of his treasure in the Dove, of which 300,000 silver coins were found in 1831, and more in 1883.

The ruins came into the possession of John of Gaunt, who built a more comfortable residence than a military castle. On his death it became Crown property and has remained so since.   Mary, Queen of Scots, was brought here several times during her imprisonment, occupying a timber building in the tiltyard.   The castle was demolished, like that of Stafford, by Sir William Brereton in 1646, after it had been held by Royalists since the beginning of the war.   The present remains are only in part genuine remnants of John of Gaunt's castle.   The " Julius Tower " is an erection of the early nineteenth century.

Chartley Castle

Sir William de Caverswall built a castle at Caverswall in 1292, but it was destroyed.   The present building is a Jacobean manor house, of unusually strong construction, which superseded an Elizabethan farmhouse on the same ground.   It was garrisoned for the Parliament during the Civil War, but apparently was never besieged.

The name Newcastle-under-Lyme was given in the

twelfth century to the building which superseded an older fortress at that place. Some say the latter was at Chesterton, two miles to the north, where some ruins could be seen as late as the sixteenth century. Opinion nowadays is in favour of Trentham having been the site of the old castle. The new stronghold was possibly of wood on a stone foundation. Hard by was the forest of Lyme.

Chartley Castle, midway between Stafford and Uttoxeter, was built by William, Earl Ferrers, about the year 1220. Judging by its ruins the fortress must have been of immense strength. Under the nunnery at Alton can be seen traces of a castle ; it was in all probability built by the same Bertram de Verdon who founded Croxden Abbey. Dr Plot, in his famous history of Staffordshire, published in 1688, mentions a Stourton Castle, and Henry de Audley owned a stronghold at Heighley in 1226. The south-west of the county was overawed or protected, as the case might be, by the castle at Dudley just over the border in Worcestershire.

The castle at Eccleshall was rebuilt by Bishop Walter de Langton, of Lichfield, in the year 1315. In 1643 it was held for the King by Bishop Robert Wright till he died, when the defence was continued by the besieged for a short time. Parliamentary forces captured it and its treasures, and rendered it uninhabitable. At the end of the century it was restored by Bishop Lloyd, and was used as the Bishop's Palace until the death of Bishop Lonsdale in 1867, when it was given up, the absence of railway communication making it unsuitable as the residence of a bishop with so large a diocese as that of Lichfield.

16.   Architecture—(c) Domestic.

Staffordshire is rich in the possession of many fine old houses.   Various mansions of the Tudor and early Stuart periods have been mentioned already in connection with the castles on whose sites they stand, but there are many more.

Wootton Lodge, between the villages of Wootton

Wootton Lodge

and Denstone, situated at the end of a beautifully wooded gorge leading up to the Weavers, is in the Elizabethan style with a front designed by Inigo Jones.   The house was built by Sir Richard Fleetwood early in the seventeenth century, and was held by him for the King until in 1643 he was obliged to surrender without resistance to a large force of the enemy.   Not far away is Wootton

Hall, built about 1730, interesting because Rousseau stayed there for some time. His favourite retreat, a cave near the house, is still to be seen.

The Oak House, West Bromwich, and Seighford Hall near Great Bridgeford, are good examples of half-

The Oak House, West Bromwich

timbered houses. Others are still to be seen at Tamworth, Leek, Stafford, Cheadle and other towns.

Ingestre Hall is a fine building of Jacobean style. The original building caught fire in 1882 and narrowly escaped complete destruction : it has been rebuilt so as to resemble the older edifice as closely as possible.

The material of which houses are built depends upon the facilities for obtaining brick or stone. Thus at Rocester the dwellings are of brick, made probably at the existing brick-works near the station; but at Hollington, Alton, and Ellastone, all within easy walking distance, stone from the local sandstone quarries is largely

Ingestre Hall

used. In the Moorlands the smallest cottages are built of squared limestone or sandstone blocks. Coal generally occurs associated with clay, so that coal-field villages and towns, and consequently the buildings of the Potteries and Black Country are constructed of uninteresting and ugly brickwork.

17. Monasticism in Staffordshire.

Staffordshire, with the rest of England, was Christian-ised by monks. St Chad, who had been trained in the monastery of Lindisfarne, and who became Bishop of the Mercians in 669, placed his seat at Lichfield and gave organisation to the great see which then included many other counties besides Staffordshire, and indeed extended from the Trent to the Thames. His episcopate marks the change from the missionary stage to that of resident clergy. He himself settled, with a band of teachers, at Chadstowe (now Stowe) near Lichfield, whence he made preaching tours up and down his great diocese, along the Roman roads which intersected near Lichfield, in the river valleys of the Trent and Sow, and across the open Moorlands to the north-west, teaching and baptising the heathen, strengthening the Christians, and establishing where possible a resident priest. Possibly we have memorials of this work in the Wells (e.g. " Chad's Well " at Stowe) which are called by his name, and which he may have used for baptisms, while the ancient dedications to St Chad at Stafford and Seighford no doubt tell of churches established by him.

At Stone another monastery was founded in these early times, and St Ermenilda, wife of King Wulfhere, also founded a nunnery there. Wulfhere's daughter, St Werbergh, founded nunneries at Trentham and Hanbury, and another Saxon saint, Modwen, had a religious house at Burton-on-Trent. Most of these were destroyed by the heathen Danes, but when the Lady

Seals of Staffordshire Monasteries

(Originals in British Museum)

1. St Mary's Priory, Tutbury (Benedictine). Reverse, 2nd seal, XV century.
2. Obverse of the same seal.
3. Croxden Abbey (Cistercian). XIV century seal.
4. St Thomas' Priory, Stafford. XV century seal.
5. St Mary's Priory, Sandwell. Late XII century seal.
6. St Mary's Priory, Rocester (Austin Canons).

Ethelfleda, daughter of the great Alfred, made effective opposition to the marauders, new centres of religious work began again to be founded. Bodies of clergy living together semi-monastically according to rule, whence termed Canons, were at Stafford, Gnosall, Penkridge, Wolverhampton, and Tettenhall. From these centres they worked in the surrounding districts, and gradually the county was won back to religion and civilisation.

A fresh revival of monasticism, too, made itself felt, and this time of a stricter type. This was the famous Benedictine rule. At Burton-on-Trent, where the great road to the north crossed the Trent, ever since Roman times a place of importance, a great Benedictine Abbey was founded in 1104. Around it the town of Burton grew up, and throughout the middle ages it maintained the great bridge which was one of the chief means of crossing the broad and often-flooded Trent—such an effectual barrier between the north and midlands of England. In the Abbey of Burton was written one of the most valuable of the monastic chronicles ; and, as we have seen, the famous protest of Archbishop Stephen Langton and the clergy of England against the Papal Legate's interference with the rights of the Church of England was received by the Legate at Burton Abbey.

Another Benedictine house was founded in the county before the Norman Conquest. This was Lapley, which was a dependent priory belonging to the Abbey of Rheims.

The Norman Conquest gave a great impetus to monasticism. In acknowledgment of the help which the Norman invaders had received from the prayers of the Norman monks, Henry de Ferrers established near his castle at Tutbury a priory dependent on the great Abbey of St Pierre-sur-Dives. Hugh, Earl of Chester, founded Trentham as a help towards re-establishing the authority he had lost in the district when the Earldom of Chester was created. Stone was refounded as an Austin Priory early in the twelfth century by Robert de Stafford with a similar object. Other Austin Priories were founded at Rocester, Calwich, Lichfield (St John's), Ronton, and Stafford (St Thomas's). The last was dedicated to the memory of Thomas Becket, on land given by a wealthy burgess, soon after the Archbishop's murder. There were Cluniac houses at Cannock and Dudley, and Benedictine nunneries at Brewood and Farewell.

The Friars reached Staffordshire in the reign of Henry III. There were Grey Friars at Lichfield and Stafford, and Black Friars at Newcastle-under-Lyme. At Radford, near Stafford, was a hospital for lepers, and a house of Austin Friars was founded at Stafford in Edward III's reign. The Knights Templars had a Preceptory at Keele.

The Cistercian Order built its houses in the waste tracts into which the other Orders had never penetrated. The Benedictines were scholars and settled near towns— the Abbot of Burton founded Burton Grammar School— but the Cistercians were agriculturists. The land they received was particularly favourable for pasturage, and the

growth of the wool trade in England made them masters
of the most profitable branch of English industry. The

Croxden Abbey

earliest house of Cistercian monks in Staffordshire was
that at Croxden (1179), in a typically lonely valley, and
the ruins of this are by far the finest and most complete

of any monastic remains now existing in the county. In 1214 the Abbey of Dieulacres, near Leek, was founded, and soon afterwards Hulton, near Newcastle-under-Lyme. All these did a great trade in wool, Staffordshire wool being then almost unexcelled in England, and many had the privilege of holding fairs and markets. But they were founded too late ever to become really great and rich houses. Laymen soon began to compete with the monks, and Kings levied severe taxes for the Hundred Years War. Unprogressive methods and indiscriminate charity brought indebtedness. Stone complained in the fourteenth century that it was impoverished by the many claims on her hospitality by travellers along the King's highway, and in the early years of Henry VI's reign Burton was absolutely insolvent and was put into commission for seven years. Disputes with neighbouring landowners and tenants became frequent. The religious houses became more and more secular in tone : little more than communities of comfortably-situated country gentlemen; no worse, but perhaps not much better than their neighbours. The monks of Rocester kept a pack of hounds. The monasteries steadily slipped into the same sort of demoralisation and decline as that which marked society as a whole in the fifteenth century, and when Henry VIII wanted money they fell almost unnoticed. In many places there was a pretence at justifying their suppression by a general allegation of abuses, but in Staffordshire such a transparent device was considered superfluous : no charges of any kind were brought against any religious house in the county.

Public opinion was largely indifferent to their fate, and even before Parliament had legalised the Dissolution of the Monasteries Calwich was quietly suppressed (1538) without remark. Their total annual revenue at the Suppression amounted to £1874, besides the valuable sites, buildings, and contents.

18. Communications—Roads, Canals, Railways.

The old roads of the county have been dealt with in the chapter on Antiquities. It was stated there that as a rule they are in use at the present time, but certain portions have been superseded by easier routes, or, having become unnecessary by the cessation of the Roman military occupation, they have become buried in vegetation and destroyed by the farmer and builder. Numerous intersecting roads have been made, and the surfaces are, on the whole, very good.

In Staffordshire we frequently come across the remains of old pack-horse tracks. In the Moorlands especially these were in use till recently for the conveyance of cheeses to market and ores from the mines. They are often found leading from one village to another; and when this occurs we may be tolerably certain that these villages at one time possessed markets. The monks used pack-horses a great deal. Salt from Stafford and Cheshire was carried in this manner, and was almost the only necessary which the medieval village could not itself produce. Hence, as

Mr Burne points out in his *Excursions into Staffordshire History*, wherever names like Saltersway, Salterswell, and Saltersford occur we may be almost sure there was a pack-horse road. The Essex Bridge and Hanging Bridge at Mayfield were probably used as pack-horse bridges. The illustration given on p. 98 shows two hollows with a mound between. These are locally termed "Cromwell's Trenches"

Essex Bridge, Great Haywood

or "Oliver's Mound," but are probably the remains of a pack-horse road. A stream runs down one of them in wet weather. This doubtless washed away the soft marl as it was stirred up and deepened the track until the wayfarers were obliged to make a fresh one. In other parts of the county, notably near Bagnall, we find a track deeply sunk between hedge banks.

Besides the roads which follow the tracks of the ancient ones, there are several old coaching roads. One of these went from Birmingham to Lichfield, then across Needwood Forest to Sudbury on its way to Ashbourne. Another coach route, from Birmingham to Holyhead, ran through Wolverhampton. Yet another main route ran across the county from south-east to north-west. Coaches on their

Oliver's Mound, Denstone

way from London to Manchester left the Watling Street near Weeford and travelled northwards to Lichfield. This divergence probably accounts for the decay of the Watling Street from this point to Muckley Corner, where it is crossed by the road from the Black Country to Lichfield. Leaving Lichfield with its numerous inns the coaches pushed forward to Rugeley and along the Vale of Trent to Newcastle and Talk-o'-th' Hill. At Stone this road was joined by another from Stafford and Wolverhampton. These coach routes, after a period of comparative neglect, are utilised now by motor cars.

Before the advent of the railway the canals were the chief means of transport for minerals and heavy goods. They are still very largely used.

The Trent and Mersey Canal connects the Mersey at Runcorn with the Trent near Castle Donington. Just before it enters Staffordshire the Macclesfield Canal joins it. The canal passes from the basin of the Mersey into that of the Trent by Harecastle Tunnel, bored through the watershed. This canal brings to the Potteries much of the material which has come by sea to the Mersey from the clay deposits of Dorset, Devon, and Cornwall. There are offshoots in the Pottery district, and at Shelton the traffic is increased by the barges which carry from Froghall the limestone quarried at Caldon Low and brought down to the water by mineral railway. Some of the limestone is used in the Potteries, but the majority is shipped along the Trent and Mersey Canal as far as Great Haywood, and thence by the Stafford and Worcester Canal into the Black Country, where a network of branches gives access to nearly all parts of the manufacturing district.

The Trent and Mersey Canal continues to follow the course of the Trent through Rugeley, and at Armitage separates from it till Alrewas is reached, after which its course is roughly parallel with the river until, leaving Burton on the right, it enters Derbyshire near Egginton.

At Fradley, near Alrewas, another canal joins the Trent and Mersey on its south side. This is a short canal formed by the junction, east of Lichfield, of the Coventry Canal, from Warwickshire, and the Wyrley and Essington Canals, which communicate with the Western canal system at Wolverhampton.

In the west of the county the Shropshire Union Railway and Canal Company owns the Liverpool and Birmingham Junction Canal, which runs north-west from Birmingham through the Black Country, and then in a fairly direct line through Brewood and Market Drayton on its way to Liverpool.

The Churnet Valley branch of the North Staffordshire Railway is laid parallel to the course of a canal which at one time linked Uttoxeter and the Potteries. In some places the line runs along the old canal bed, but in others there is a little water left between the banks even now; otherwise the bed is dry.

Barges drawn by horses travel on these canals at the rate of two to three miles an hour. Small tugs can be used which increase the pace by several miles per hour, but the limit is soon reached, for at high speed the wash from the bows causes damage to the banks, unless these are faced with concrete at great cost. Canals are useless therefore for the quick delivery of goods, but as their working expenses are less than those of railways, they are utilised for carrying non-perishable goods at a cheap rate. Unless it is found possible in the future to make the charges for carriage by rail much less than they are at present, canals will continue, at any rate in Staffordshire, to be an important means of transport for minerals and loads of a similar nature.

In 1845, except for one or two short mineral lines, there were only two railroads in Staffordshire. One was the Grand Junction Railroad, which connected Birmingham and Liverpool. This and the London and

A Black Country Waterway

Birmingham Railroad were amalgamated, and now form a part of the London and North Western Railway Company's system. The other line, the Birmingham and Derby Railroad, entered Staffordshire near Tamworth, and left it again near the confluence of the Dove and Trent. This is nowadays a part of the western main line of the Midland Railway.

Since 1845 the North Staffordshire Railway has been built. It has numerous branches in various parts of the county, and does nearly all the carrying trade of the Potteries. The London and North Western Railway has extended its system; the Great Western has some branches in the south; and the Great Northern runs trains from Derby and Grantham over the North Stafford metals to Uttoxeter, and thence by its own line to Stafford.

19. Administration and Divisions.

The Saxon tribes brought over with them and established in this country their methods of local government. These were of an extremely democratic nature. Every township had its own local council which every freeman had the right to attend. It was presided over by an officer called a "reeve" and settled all minor matters. Affairs beyond its powers were transferred to a more important court, the Hundred-moot. The Hundreds were districts so called, some think, because they provided a hundred soldiers; others are of the opinion that they were the areas occupied and cultivated by a hundred families. Each town moot sent its reeve and four delegates to help form the Hundred-moot, which was summoned by the "Hundred-man." On this council sat the Eorls and Thegns (or Thanes) who lived in the district and were land-owners and officials of the king.

There were five Hundreds in Staffordshire, each divided now into two. *Pirehill Hundred*, North and

South, takes its name from a hill overlooking the Trent a short distance south of Stone. *Cuttlestone Hundred*, East and West, is perhaps called after Cuttlestone Bridge, at Penkridge: the clan probably met at the Stone of the Gorsedd, or Judgment-seat, at Penkridge. *Seisdon*, a little place south of Wolverhampton, in Trysull, was formerly a place of importance and gave its name to another Hundred, now also divided into two sections, North and South. *Totmanslow Hundred* derives its name from the place of the same name near Tean. Some say the name means the low, or burial-mound, of Tatman, the "bright and happy fellow": on the other hand it is also thought that it was associated with the worship of the Celtic deity Toutates. North-east Staffordshire comprises the North and South Divisions of this Hundred. *Offlow Hundred*, in its two divisions, includes the east and south-east of the county, and takes its name from a burial mound near Lichfield where, very probably, the Hundred-moot used to meet.

Formerly, if a crime was committed, such as robbery, arson, killing or maiming cattle, destroying roads, etc. the Hundred had to make it good, but in the reign of George IV an Act was passed leaving the Hundreds liable only for damages done by rioters. Even that liability is now removed and the old divisions are used only by the magistrates as boundaries of the petty sessional divisions. The boroughs of Stafford, Newcastle-under-Lyme, Tamworth, Walsall, Wolverhampton, and the city and county of Lichfield form modern Hundreds. Disputes between people living in different Hundreds were settled by the

Folk-moot which, when grouping into shires took place, became the Shire-moot, which corresponded in some respects to the modern County Council, and was presided over by a Shire-reeve or Sheriff, the direct representative of the King. The Sheriff collected the taxes for the King and took them to the Court of Exchequer in London together with an accurate statement of accounts, which was carefully examined, and all matters of dispute were settled by this Court, which was originally a special meeting of the King's Privy Council. The Sheriff attended to other royal business in the shire and was responsible for seeing to the carrying-out of the Shire-moot's decisions. The Town-moot system gradually came into the hands of the lord of the "Manor"—which was frequently co-terminate with the township. Then under the feudal system the Government passed more and more into the hands of the King, his centralised law-courts, and the Circuit Judges.

The numerous wars in which this country was involved at different times often took its King and his officers abroad. This resulted in the imperfect maintenance of law and order in England, partly no doubt owing to the return of time-expired soldiers of no habits of industry and no occupations to a necessarily disturbed country, and partly to the fact that the riff-raff did not take part in the wars and stayed behind with fewer honest men to look after them.

The narration (from *English Wayfaring Life*, by Jusserand) of an occurrence in 1342 during Edward III's bellicose reign will show the insecurity of the high roads at that time.   " Some Lichfield merchants state to their lord, the Earl of Arundel, that on a certain Friday they sent two servants and two horses laden ' with spicery and mercery,' worth forty pounds, to Stafford for the next market day.   When their men ' came beneath Cannock Wood ' they met Sir Robert de Rideware, Knight, who was waiting for them, together with two of his squires, who seized on the domestics, horses, and booty, and took them to the priory of Lappeley.   Unfortunately for him, during the journey one of the servants escaped.   At the priory the band found ' Sir John de Oddyngesles, Esmon de Oddyngesles, and several others, knights as well as others.'   It was evidently an arranged affair, carefully organised ; everything was done according to rule ; they shared ' among them all the aforesaid mercery and spicery, each one a portion according to his degree.'   That done, the company left Lappeley and rode to the priory of Blythebury, occupied by nuns.   Sir Robert declared that they were the king's men, ' having travelled far,' and begged for hospitality as it was usually given.   But it seems the company had a bad appearance ; the abbess refused.   The knights, seeing this inopportune reception, burst the gates of the barns, gave hay and oats to their horses, and so passed the night.

But they were not the only persons well occupying their time.   The escaped servant had followed them at a distance, and when he saw them installed at the priory he returned in all haste to Lichfield to warn the bailiff who hastened to collect his men for the pursuit of the robbers.   The latter, who were men of the sword, as soon as they were met, turned on their defence, and a real combat took place, in which at first they had the upper hand, and wounded several of their enemies. At length, however, they lost ground and fled; all the spices were recovered, and four of their company taken, who were immediately beheaded on the spot.

Robert de Rideware was not among the victims, and did not lose heart. He met his relative Walter de Rideware, with some of his following, while the bailiff was on his road back to Lichfield; all together turned their horses' heads in pursuit of the bailiff. A fresh fight; this time the king's officer was worsted and fled, while the lords finally took from him the spices once more.

What resource remained for the unhappy William and Richard, authors of the petition? Resort to justice? This they wished to do. But as they were going for this purpose to Stafford, chief town of the county, they found at the gates of the city some of the retainers of their persecutors, who barred their passage and even attacked them so warmly that they hardly escaped without grievous hurt. They returned to Lichfield, watched by their enemies, and led a pitiable existence. 'And, sire, the aforesaid William and Richard, and many people of the town of Lichfield, are menaced by the said robbers and their maintainers, so that they dare not go out of the said town at all.' This document, the original of which still exists, is characteristic of the state of justice at that period."

Nowadays "summary jurisdiction" for minor offences is exercised in petty sessional courts. These are held by justices of the peace, commonly called magistrates, who are properly qualified men recommended to the Crown by the Lord-lieutenant of the county and are appointed on the King's behalf by the Lord Chancellor.   The list of magistrates for a district is known as its Commission of the Peace.   Two or more justices of the peace sitting together, or a stipendiary magistrate, constitute a Court of Petty Sessions.   Staffordshire is divided into twenty-four petty sessional divisions for the country districts, with additional courts and Commissions of the Peace for the larger towns.

In addition to these minor courts there are Courts of Quarter Sessions, held four times a year in counties and certain boroughs.   In Staffordshire there is one court, held at the county town.   Very little remains of the civil jurisdiction of quarter sessions except as to orders for closing and diverting highways, certain jurisdiction as to lunatics, and the appointment of visiting committees for prisons.   They have jurisdiction in burglary cases but these are usually tried at assizes, that is, the periodical courts held by the Judges of the King's Bench Division. The country is divided into certain districts known as Circuits, each with its own group of judges and officials. Staffordshire is in the Oxford Circuit.   The serious cases, with certain exceptions which are dealt with in special London courts, are tried at the assize held next after the preliminary enquiry by the magistrates.

Special sessions are held by the justices of a division for certain purposes, such as licensing or brewster sessions, sessions for hearing appeals against parish rates, for the appointment of constables, and other purposes.

The control of the police is in the hands of a Standing Committee appointed jointly by the Quarter Sessions and the County Council. Some County Boroughs have their own police forces.

The Lord-lieutenant of a county is the permanent

Town Hall, Burslem

local representative of the sovereign in the county, is the medium of communication between the government and the magistracy, and is intrusted with the maintenance of the public peace. He presents to the sovereign the names of county gentlemen to be appointed deputy-lieutenants. Justices of the peace are sub-deputy lieutenants. Lord-lieutenants are appointed by the Crown by patent under the great seal and hold office for life or during good behaviour.

As has been seen already, the office of sheriff in an English county is of great antiquity. Since 1340 sheriffs have been appointed annually. With the gradual trans-ference of legal matters into the hands of the King and his judges, and the collection of taxes to officers of the great revenue departments, the sheriff's duties have altered, and in 1887 the law concerning them was consolidated. The Lord Chancellor, the Chancellor of the Exchequer, members of the Privy Council, and the judges meet at the Law Courts on the 12th of November and select three names for each county after hearing claims for exemption. These names are submitted to the King, who pricks a hole in the parchment opposite one name (generally the first) for each county. In the counties outside the Central Criminal Court district the sheriff attends the judges when on circuit in his county, and he is the returning officer at parliamentary elections. The other duties of the office, which are connected with the summoning of juries, the execution of civil process, the hearing of compensation cases, and attendance at the execution of criminals, are usually performed by the under-sheriff or the deputy-sheriff. The deputy-sheriff, appointed by the sheriff, must have an office in London for the receipt of writs of the High Court. In cities and towns which are counties of themselves (such as Lichfield) the sheriff is appointed annually by the town council on November 9, immediately after the election of the mayor.

After Saxon times until recently the tendency was for all power to be centralised, so that the extent to which a county governed itself became less and less. In modern times the tendency is the reverse.

Staffordshire is divided into fifteen poor-law unions wholly within the county area and seven partly within its boundaries. The law concerning the relief of the poor, both the able-bodied idlers and the genuine poor, is the development of the earliest legislation passed in 1601. Before that time the monasteries did what they could in this respect, but their charity was indiscriminate, and, like the giving of money to tramps with plausible tales, encouraged mendicancy. Nowadays there are two kinds of relief, indoor and outdoor. The latter is given to the infirm, the aged, and others in their own homes: the former refers to the maintenance of the poor in work-houses. The able-bodied pauper must complete a definite task in return for food and lodging. Poor-law unions sometimes extend into two or more counties. This is the case with several Staffordshire unions.

Parish Councils, established in 1894, have certain powers relating to the appointment of overseers who assist in the administration of the poor-law; the provision of a parish hall, recreation ground and allotments; the water supply and sanitary arrangements; and certain other matters of purely parochial interest. They cannot raise more than a threepenny rate without the consent of a meeting of the parliamentary or county electors of the parish.

County Councils were created in 1888, to fill the void left by the disappearance of the ancient shire-moot. They are elected bodies, consisting of chairman, aldermen, and councillors, and have very wide duties. The county council is the authority to fix the county rate ; to maintain the county buildings, main roads, bridges, and lunatic asylums; to appoint and pay coroners, public analysts, and other officers; to carry out numerous acts relating to public health, weights and measures, and diseases of animals. It has numerous other powers and duties, not the least important being the control by a special committee of much of the elementary and higher education of the county.

The recent acts relating to Old Age Pensions and National Insurance are administered by special bodies.

County Boroughs, first created in 1888, combine the powers of an Urban District with those of a County Council. They are independent of the County authority in almost every point, being entirely self-contained for all rating purposes, public health, education, etc. The County Boroughs in Staffordshire are Burton, Smethwick, Stoke-on-Trent, Walsall, West Bromwich and Wolverhampton.

Local bodies are controlled by the Local Government Board in the matter of raising loans.

Staffordshire is entirely in the diocese of Lichfield in the province of Canterbury. It is divided into the archdeaconries of Stafford and Stoke-on-Trent and these again into rural deaneries. The various religious bodies outside the Church of England have their own systems of government.

Staffordshire has seven Parliamentary divisions, each sending one representative to the House of Commons: these are the North-Western, Western, Burton, Handsworth, Kingswinford, Leek, and Lichfield divisions. There are also Borough Members for Hanley, Stafford, Stoke, Walsall, Wednesbury, West Bromwich, and Wolverhampton.

20. Roll of Honour.

St Chad came from the north to be Bishop of Mercia at the request of Wulfhere, King of Mercia, who founded a monastery at Stone. He died in 672 and his copy of the Gospels is still kept in the Lady Chapel at Lichfield. Bishop Hedda's church of St Peter, at Lichfield, remained the Cathedral until it was replaced four hundred years afterwards by Bishop Roger de Clinton's Norman structure. Under Offa's rule there were two Archbishops of Lichfield, Higbert and Aldulf.

Ethelfleda freed East Staffordshire from the grip of the Danes, and fortified Tamworth and Stafford (where St Bertillinus had lived as a hermit). She made Gnosall into a Royal Free Chapel, where neither sheriff nor bishop had jurisdiction, and Edgar did the same for Tettenhall.

Though Staffordshire seems remote from the busiest scenes of English history, yet the vicissitudes of its principal families show that almost every member was actively concerned with the political events of his time. To give the deeds of the Staffords, the Ferrers, the Audleys and the Dudleys who had distinguished themselves would be to write a large portion of the history of England.

The most famous of the Pagets is the "Waterloo Marquess," who at this battle led the great cavalry charge of the British centre. The Marquess of Anglesey had previously distinguished himself in Holland and during the Coruña campaign, and Lord George Paget took part in the Charge of the Light Brigade.

Certain bishops of Lichfield also held the office of Lord-Marcher, their duties being the organisation of counties in Wales and the maintenance of the security of the border. One of them was Bishop Smyth, re-founder of Brasenose College, Oxford, and also St John's College, Lichfield. According to Leland, "One Langton, Bishop of Lichfield, made the fair palace and the close waulle," and he also "made Ekleshaul Castel, Shoeborrow Maner Place, and the Palace by Stroude." This bishop (1296–1321) designed the Lady Chapel.

Mary Queen of Scots' treasonable correspondence was discovered by a spy named Gilbert Giffard, who belonged to the same family as Bonaventure Giffard, whom James II appointed as President of Magdalen College, Oxford, in disregard of the Fellows' election. When Charles I set up his standard at Nottingham, the aged head of the Giffard family garrisoned Chillington for the King, and after the battle of Worcester Mr Charles Giffard guided Charles II to Whiteladies, where he was disguised by the Penderels and taken to Boscobel, a house built by Mr John Giffard. The King escaped south by acting as manservant to Miss Jane Lane of Bentley Hall, one of whose ancestors was the first Governor of Virginia.

Conspicuous among the Parliamentarians was Thomas Harrison of Newcastle-under-Lyme. Of him Cromwell remarked, " Harrison is an honest man and aims at good things, yet from the impatience of his spirit will not wait the Lord's leisure, but hurries me on to that which he and all men will have cause to repent." As President of the Council of Thirteen he was influential in causing the Barebones Parliament to be summoned. He was described as being at that time " second in the nation."

In 1682 Jonathan Wild was born at Wolverhampton. Luckily for Staffordshire he fulfilled his career as a master-thief in London. He spent half his time in assigning work to his bands of specialists, and during the other half restored the stolen articles at a commission through his " lost " property office.

After the Restoration, Sheldon, a native of Ellastone, was made Bishop of London. He had been correctly marked out from boyhood as a future Archbishop of Canterbury. When building the Sheldonian theatre he discovered Wren's talents. One of Sheldon's acquaintances was Charles Cotton, friend of Izaak Walton and translator of Montaigne, who was born in 1630 on his father's estate, Beresford. His forebears of Beresford had been master-foresters to the King. He is perhaps best known by his association with Izaak Walton, to whose great work he added *Instructions how to Angle for a trout or grayling in a Clear Stream*. In the *Retirement* he gives

Izaak Walton

an idea of their enjoyment when visiting their fishing cottage on the Dove.

Izaak Walton himself was born at Stafford in 1593 and was at one time a linen-draper in London. He published his first edition of the *Compleat Angler* in 1653 and over 120 have since been given to the world. The work has lost no whit of its charm even in this common-place and non-idyllic age, and his *Lives* of Donne, Wotton, Hooker, and George Herbert are no less delightful. He died at Winchester in 1683.

Samuel Johnson

In 1717 Samuel Johnson entered Lichfield Grammar School. The circumstances of his early life are well known—how his studies progressed in spite of his delicacy and how he preferred reading his father's books to selling them in the market place, for which act of filial disobedience he did penance in Uttoxeter. One of his pupils at Edial Academy was David Garrick, and the two set out for London in company to seek their fortunes. It was at Lichfield Palace that Garrick began his stage

Uttoxeter Conduit, with bas-relief of Dr Johnson's penance

career by acting in *The Recruiting Officer*. He was at home in every part, whether that of Hamlet or of Lord Ogleby, and Pope said of him " That young man never had his equal as an actor, and he will never have

a rival." The "Little Mimic" brought out Johnson's
tragedy *Irene*, and though the social ideas of the two
friends were so different, yet their intercourse did not
cease until Garrick's death "eclipsed the gaiety of nations
and impoverished the public stock of harmless pleasure."
In Johnson's *Dictionary* there is a gracious reference to
his native city, where his mother continued to live on
slender means, and it was on her account that he wrote
Rasselas in the short time of a week.

Joseph Addison, though not a Staffordshire man, was
also for a short time at Lichfield Grammar School.

Another learned native of Lichfield was Elias Ashmole
(1617–1692). He ranged over many subjects, including
alchemy, and he professed to have the secret of the
philosopher's stone, but his chief delight lay in antiquarian
research, and in 1672 he published *Institutions, Laws, and
Ceremonies of the Most Noble Order of the Garter*. The
diary kept by this virtuoso tells how his collections filled
twelve waggons when he presented it to Oxford University,
thus founding the Ashmolean Museum.

John Lightfoot, the Hebrew scholar (1602–1675),
born at Stoke, was the son of the Vicar of Uttoxeter,
and became Master of St Catharine's, Cambridge.

In the latter part of the eighteenth century, the trade
of the county made great strides. "Schemer Brindley,"
the Derbyshire engineer, increased its prosperity by
making it easy for goods to be carried to and from the
Potteries and Black Country. His first task was to set
right the works of a paper-mill near Dane Bridge and
then he went on to the laying of the Bridgewater, Trent

and Mersey, Staffordshire and Worcester Canals, and to other triumphs.

The "Father of the South Staffordshire iron trade" was Wilkinson, who succeeded in substituting mineral coal for wood-charcoal in smelting and puddling. His improved boring appliances were used in the construction of cannon for the Peninsular War. He built iron barges, and took out a patent for lead pipe making. At Bradley he had coke ovens and turned out all kinds of iron-work, including a powerful steam-engine for the Paris water-works.

After the dissolution of the monasteries, the Adams family continued the Hulton Abbey pottery works. Palmer discovered the process of salt-glazing through a servant allowing a pot of pickles to boil over, and he and Adams produced Crouch ware. Astbury was the first to import white clays from Dorset and Devon, and the first to go to London to dispose of his goods. He mixed the clay with sand, but Heath introduced the use of flint for this purpose. About this time the Elersers were revealing their secrets of fine workmanship and making their wonderful red ware. Aaron Wood was apprenticed to Dr Thomas Wedgwood and acted as modeller for all the manufacturers. Ralph Shaw was prominent among the white stone salt-glaze manufacturers, for he used to make "chocolate ware" and remove its white coating in patterns. The first porcelain was manufactured by Littler. William Adams (1740) encouraged some Dutchmen in their enamelling work. They had been brought over by Warburton and Ralph Daniel, who was the first to use plaster moulds. The best work was then being carried out by Whieldon, famous for his agate, tortoise-shell, cauliflower, and pine-apple wares. Spode was his apprentice, and introduced blue printed ware. One of the Adamses was very successful in transfer painting.

Josiah Wedgwood

The younger Spode fixed the composition of the English China body.

Whieldon took as his partner Josiah Wedgwood, who began life as a "thrower." He eventually set up in business at Burslem, working at the improvement of the cream-colour ware which superseded salt-glaze pottery. It was called Queen's ware after Queen Charlotte, whose patronage he won in London where his enterprise had led him to establish show-rooms. He called his new works "Etruria," under the mistaken impression that some ware ornamented with classical designs which he much admired was of Etruscan workmanship. Wedgwood much improved the black "Egyptian ware," but concentrated upon the production of jasper ware of his own invention, in which he reproduced the famous Portland Vase. His companion in clay hunting was Adams. These two, with Turner, employed the cleverest designers of the day including Flaxman. Nor did Josiah Wedgwood's interests, or those of his family, limit themselves to his trade. He concerned himself in improving the local means of communication, Ralph Wedgwood worked at the invention of the electric telegraph, and Thomas Wedgwood was the first to try to utilise the knowledge that light affects nitrate and chloride of silver, and so to make photographs. Unfortunately his pictures took too long in the exposure, and he could not fix them.

Mason's ironstone china was invented by the father of George Mason the artist, who came home after painting pictures of the Campagna, and settled at Wetley Abbey. His landscapes reveal his appreciation of the soft colouring and beauty of Staffordshire. A far greater name in the world of art, however, is that of Peter de Wint (1784–1849) who was born at Stone, and will always be held one of the very finest of English water-colourists.

The charm of Staffordshire appealed to two poetesses

of the nineteenth century—Miss Dakeyne, who wrote *Legends of the Moorlands and Forest*, and Mrs Cantrell, authoress of *Melodies from the Mountains.*

Admiral Lord Anson

Dinah Maria Mulock, afterwards Mrs Craik, was born at Stoke-on-Trent. Her first three-volume novel, *The Ogilvies*, was very popular, but her best-known work is *John Halifax, Gentleman.* Besides writing many pleasant

tales for the young she gave some time to poetry, and published in 1881 a collection called *Poems of Thirty Years*.

Admiral Lord Anson was born at Shugborough, and
joined the navy in 1712. As commodore he commanded
a squadron of six ships sent to attack Spanish possessions
in South America. After taking Paita in 1742 he was left
with only one, the *Centurion*, but managed to capture the
Manila galleon, said to be the richest prize ever taken.
When he was Rear-admiral of the White the condition
of the fleet was very poor and in the far-reaching reforms he
effected lies one of his claims to renown. In 1747 he commanded the Channel fleet in an action off Cape Finistère
against a French convoy, for which he received his title.

Lord Anson was distantly related to Admiral Jervis.
From the standing of able seaman Jervis soon made his
way, till in 1759, as acting commander of the *Porcupine*,
he had the task of leading the advanced squadron in
charge of the transports past Quebec. On this occasion
his prompt decision roused the admiration of General
Wolfe. Newfoundland was recovered after some operations in which he took part, and later he became Com-
mander-in-Chief in the Mediterranean. In 1797 he
posted himself off St Vincent so that no ships could come
to the support of the French, and on the 14th of February
engaged the 27 ships. He broke up the Spanish fleet
with his 15, inflicting a signal defeat and thus relieving
England from the fear of invasion. For this he gained
his title of Earl of St Vincent.

Sir Harry Smith Parkes upheld England's honour in
the East. He was born at Birchills Hall in 1828. Early

in life he went to China, where as consul his determination stood him in good stead during the wars. For three years he governed Canton, and for eighteen more was minister in Japan, where he supported the reform party.

The self-sacrificing devotion of Dorothy Pattison, better known as "Sister Dora," to hospital work in the latter part of the nineteenth century is commemorated by a statue at Walsall.

21. THE CHIEF TOWNS AND VILLAGES OF STAFFORDSHIRE.

(The figures in brackets after each name refer in nearly all cases to the population of the civil parish in 1911. Some villages are included solely on account of their possession of some interesting feature, while places of less than 2000 people are omitted if they are of no special interest. The figures after each paragraph refer to the pages in the text.)

Abbots Bromley (1467), formerly a market town, is situated between Lichfield (11 miles) and Uttoxeter (7 miles). It is famous for its September wake, with hobby-horse and reindeer-horn dances (see p. 49). The church is in the Early Decorated style. The old market cross still remains.

Acton Trussell (523), on the river Penk, four miles from Stafford. The Early Decorated church has some good stained glass and was formerly attached to Stafford Priory. With Acton is associated Bednall.

Aldridge (2812), about four miles north-east of Walsall, has a church of Perpendicular and Decorated periods, containing two ancient effigies, one dating from the time of Henry III and the other a Crusader in chain armour. There are important brick and tile works and some collieries.

Alrewas (1461), a low-lying village on the Trent, five miles north-east of Lichfield. The name (pronounced all-roo-as) means "the alder swamp." The church, founded in 820 A.D., is in Norman, Decorated, and Perpendicular styles, with various points of interest. Baskets are made here. (p. 99.)

Alstonfield (457), one of the wildest and most picturesque parishes in the county, lying to the west of Dovedale. An old font, a stone coffin and parts of some early crosses are in the churchyard. (p. 69.)

Alton (1279), is a village on the Churnet Valley line of the North Staffordshire Railway. Stone-built, nestling at the foot of

The Gardens, Alton Towers

precipitous heights clothed with pine-woods, and crowned with castle and convent with towers and turrets, its aspect is quite un-English. This is in the Rhineland of Staffordshire. The castle, now in ruins, was built in 1175. Pugin designed the chapel and convent, which date from 1840. There are important local sandstone quarries. Alton Towers, with its wonderful landscape gardens, the seat of the Earl of Shrewsbury, is in the parish. (pp. 4, 13, 57, 71, 86, 89.)

Armitage with Handsacre (1565), three miles south-east of Rugeley. The cast-iron bridge spans the Trent in a single arch of 140 feet, one of the longest in Britain. At Hawkesyard is a Dominican Priory with a school and college, with power from the Pope to confer University degrees in Philosophy and Theology. (p. 99.)

Ashley (756), is a straggling village seven miles from Eccleshall, and has an interesting church with ancient monuments, and some sculpture by Chantrey. (p. 48.)

Audley (14,776), five miles north-west of Newcastle-under-Lyme, has coal and iron mines. In the church is a brass dating from 1385 with an inscription in Norman French. There are also two effigies. Three miles away, to the south-west, the Lords of Audley built Heighley Castle.

Bagnall (662), five miles from Stoke, is said to have been a market town in Saxon times. The cross still remains. (p. 97.)

Barton-under-Needwood (1554), between Burton and Lichfield, has cement works and had formerly a fair.

Baswich (849), a composite civil parish two miles south-east of Stafford. There are salt works, and the pumping station for Stafford Waterworks is in Milford. (p. 58.)

Betley (761), seven miles north-west of Newcastle-under-Lyme, on the borders of Cheshire, has a Perpendicular church with pillars formed from single tree-trunks.

Biddulph (7422), three miles south-east of Congleton, eight miles from Leek. The church has brasses and an interesting font. On the moor the river Trent rises. There are ironworks, collieries, quarries, and cloth factories. The grounds of Biddulph Grange contain many botanical curiosities. (pp. 56, 71.)

Bilston (25,681) is the Billestune of Domesday, and in a charter of 996 A.D. is called Bilsretaune. The working of iron and other metals is the chief industry. (pp. 53, 55, 58.)

Black Heath (6679) is a part of Rowley Regis.

Blore (144), four miles north-west of Ashbourne, is above Dovedale. In the church are effigies of knights in armour, a canopied tomb, and old stained glass.

Bloxwich (8411), a busy mining and manufacturing centre three miles north-west of Walsall, of which it is a ward; makes locks, buckles, bits and stirrups, needles, and awl blades. (p. 55.)

Brereton (2675) near Rugeley, well known on account of its collieries.

Brewood (2567), eight miles north of Wolverhampton, is an ancient town, formerly possessing a market and a residence of the bishops of Lichfield. The Giffard family has lived here since the time of Henry II. (pp. 93, 100.)

Brierley Hill (12,263) is a market town manufacturing glass, chains, boilers, spades, and fire-clay products.

Broughton, about five miles from Eccleshall, has in its church three ancient windows. The Hall is a good example of Elizabethan domestic architecture. (p. 80.)

Brownhills, see Ogley Hay and Norton Canes.

Burntwood (8636), four miles west of Lichfield, has in its parish a house formerly occupied by Dr Johnson. The County Lunatic Asylum, with accommodation for over 800 patients, is situated here.

Burslem (41,566), called Barcardeslim in Domesday Book. The "Mother of the Potteries" has been for several centuries famous for its earthenware. The museum contains examples of the work of Josiah Wedgwood. There are ironworks and other industries. It became part of Stoke-on-Trent in 1910. (pp. 59, 60, 62, 108, 120.)

Burton-upon-Trent (48,266). The river Trent is navigable for barges up to this place. Burton is the largest brewing centre of England, and the enormous stacks of beer-barrels proclaim its chief industry. Railways cross the town in all directions, serving the various brewing premises. Burton beer is said to owe its special quality to the water, which contains calcium sulphate, gypsum being quarried in the neighbourhood. There were important monastic institutions here, but the remains are scanty. British, Saxon, and Roman relics have been found in this district and may be seen in the museum. The population of the town has decreased by 2120 since the 1901 census. (pp. 6, 8, 15, 52, 57, 63, 65, 67, 72, 90, 92, 93, 95, 99, 111, 112.)

Bushbury (3594), just north of Wolverhampton, has an old church in several styles, including Norman.

Cannock (28,586 in the Urban District), nine miles from Stafford and ten west of Lichfield. No one would imagine that Cannock was once a watering-place of repute, for it resembles a " Black Country " district to-day. It was called Chnoc in 1130 and Cank in the fifteenth century. The "Chase," which takes its name from Cannock, was once well-wooded. Though still wild in parts, it is dotted here and there with collieries and ugly mining villages. (pp. 14, 56, 58, 93.)

Castlechurch (1660), two miles from Stafford. The name explains itself. There are no remains of the fortress, and only the tower of the church is old.

Cauldon or Caldon (483), eight miles south-east of Leek, is noted for its limestone quarry, which has eaten away a great part of the "low" or hill in the course of several centuries' working. Flint implements have been found here. (pp. 57, 99.)

Caverswall (5264), three miles from Longton. A castle was built here in Edward II's reign and was rebuilt in the time of James I: it is now a private residence. (p. 85.)

S. S.   9

Chartley Holme (49), eight miles from Stafford, on the Uttoxeter line, was famous for its herd of wild white cattle. There are some remains of the ancient castle. (pp. 37, 65, 67, 86.)

Chase Town and **Chase Terrace** (5514), four to five miles from Lichfield, are two large mining villages. The parish church of St Anne's was the first church in England to be lighted by electricity.

Cheadle (5841), a market town 13 miles north-east of Stafford. There are collieries, a tape factory and other industrial works here. (pp. 43, 52, 56, 58, 88.)

Checkley (2280), about four miles south-east of Cheadle, has an interesting church, partly Norman but chiefly of later date. There is a rude Norman font and an effigy of a Norman crusading knight. In the churchyard are the remains of an ancient cross. (pp. 74, 80.)

Cheddleton (3221), three miles south-west of Leek, stands in a pleasant position above the Churnet valley. The church is interesting and there is an old cross in the churchyard. Pottery and tissue paper are made here.

Chesterton (7145), two miles from Tunstall and New-castle, reflects its Roman origin in its present name. It is probably the "Mediolanum" of the Antonine Itinerary. (pp. 72, 74, 86.)

Church Eaton (593), six miles south-west of Stafford. The church has an unusual dedication to St Editha, and is of very ancient foundation. St Editha's Well is close by at High Onn. (p. 78.)

Clifton Campville (460), on the river Mease six miles north-east of Tamworth, has a brass dating from about 1350 A.D. The church (Early English and Decorated) is a noble structure. (p. 80.)

Codsall (1634), 4½ miles north-west of Wolverhampton. The church has a tomb with effigy in armour, also a fine Norman doorway.

Coseley (22,834), three miles south-east of Wolverhampton, has coal mines, cement works, iron foundries, etc.

Cradley Heath (10,101), a large industrial village with chain-making works, etc., a couple of miles north of Stourbridge. (p. 55.)

Croxden (179), 5½ miles north-west of Uttoxeter, has the ruins of a once beautiful Cistercian Abbey. (pp. 78, 86, 94.)

Croxton, four miles from Eccleshall, is at the water-parting between the Humber and Severn tributaries.

Darlaston (17,107), a "Black Country" town. The principal articles of manufacture are bolts and nuts, screws, gunlocks, roofs, railway ironworks, bridges and girders, etc. Valuable iron ore is mined.

Denstone (724) is a small hamlet five miles north of Uttoxeter. The church is modern. Denstone College stands on a hill facing the Weaver Hills, in what is perhaps the finest situation for a school in England. Opened in 1873, and dedicated to St Chad, it has grown steadily, and a sum considerably exceeding £100,000 has been expended on buildings alone. In September 1914 there were 296 boys in the College and the attached Preparatory School. (pp. 47, 87, 98.)

Dilhorne (657), 2½ miles west of Cheadle, has a church containing twelfth century work. The tower is octagonal.

Draycott in the Moors (368) is a village 2½ miles south-west of Cheadle. The church, of the Decorated period, contains some fine tombs. Totmonslow, the burial mound of

Denstone College

Totman, the "bright and happy man," is half-a-mile to the east. It gives its name to one of the hundreds of Staffordshire.

Eccleshall (3683), a small and ancient market town, consists of one long street. Its nearest station is Norton Bridge, three miles distant. The church (Early English and Perpendicular) is very fine. It was for several hundred years the residence of the bishops of Lichfield. (pp. 67, 86, 113.)

Ellastone (240). The parish includes Calwich, Ramsor, Stanton and Wootton (total population 941) and is on the eastern slope of the Weaver Hills, overlooking the Dove. The district and people have been well described by George Eliot: for "Heyslope" in *Adam Bede* refers to the village; "Norbourne" is the neighbouring Derbyshire hamlet of Norbury; "Oakbourne" stands for Ashbourne, "Eagledale" for Dovedale, "Loamshire" for Staffordshire, and "Stonyshire" for Derbyshire. The "Donnithorne Arms" of the book is the Bromley Arms inn. George Eliot's own father was the original of Adam Bede and was born at Roston, near Norbury. Archbishop Sheldon was born at Stanton. (pp. 89, 114.)

Endon (1583), seven miles north-east of Stoke, has an Early Decorated church, recently enlarged. The custom of "dressing the well" with flowers is kept up on May 29th, to commemorate the Restoration. The day is kept as a holiday, with a special service in the church followed by maypole dancing on the village green.

Enville (712), a parish near Stourbridge, has an ancient church of red sandstone in Norman, Early English, and later styles, with fine tombs and carved *misericorde*. Two effigies are probably of pre-Norman workmanship. The tower is an imitation of the Somersetshire towers.

Etruria, a village in the municipal borough of Stoke-on-Trent. It was lighted with gas as early as 1826. The famous pottery works founded by Josiah Wedgwood are still carried on by his direct descendants. The King, as Duke of Lancaster, is Lord of the Manor. Etruria Hall, where Wedgwood died, now forms an iron-company's offices. (pp. 60, 121.)

Ettingshall (6643), a "Black country" village two miles south-east of Wolverhampton. The chief trade is in iron and coal; there are also flour mills.

Fazeley (1830), near Tamworth, has tape manufactures, bleaching and cotton dyeworks, and two breweries. (pp. 58, 72.)

Fenton (25,626), part of Stoke; a pottery and coal mining town. (p. 62.)

Forton (497) is a little village 1½ miles north-east of Newport on the Shropshire border. "The Monument" is a circular red sandstone building with conical roof, placed on a hill overlooking the village to commemorate something now forgotten. Aqualate Hall stands in a deer park near a large lake or mere; close by is a Roman well. (p. 48.)

Gailey and **Hatherton** form one parish about eight miles south of Stafford. Near Gailey are three large canal reservoirs which afford excellent fishing and are frequented by many water birds.

Gnosall (2067), six miles south-west of Stafford, is of note only on account of the cruciform church of St Lawrence, one of the finest ecclesiastical buildings in Staffordshire. It contains fine Norman, Early English, Decorated, Perpendicular, and some ugly modern work.

Goldenhill (4896), just north of Tunstall, is a modern pottery village, now part of Stoke-on-Trent.

Gornal. See under Sedgley.

Great Haywood (571) is one mile from Colwich station. Here may be seen the 14 arches which remain out of the original 42 of the " Essex Bridge," built by the Earls of Essex to afford an easy passage from Chartley to the hunting grounds of Cannock Chase. Shugborough Hall, near by, was the birthplace of Admiral Anson. There are some quaint Chinese and other buildings in the park. (pp. 22, 97, 99.)

Hanbury (523), six miles north-west of Burton-on-Trent. The old parish includes a large portion of the Needwood Forest. Ethelred, King of Mercia, founded a nunnery here in 680, his niece Werberga being prioress. (p. 90.)

Hanley. See Stoke.

Heath Town (12,276) near Wolverhampton, manufactures iron goods. There are coal and ironstone pits.

Hednesford (11,473) is a Cannock Chase town, populous on account of its coal pits. It is in the Cannock Urban District.

Himley (307), six miles south of Wolverhampton, has a Georgian church and a hall, a residence of the Earls of Dudley. Holbeach House, close to Himley, stands on the site of the house of Stephen Littleton, a Gunpowder Plot conspirator. It was hither that the conspirators fled. The pursuers set the house on fire, and as those inside ran out they were shot. Those who had taken refuge in the woods were captured. (p. 67.)

Hollington is a long straggling village in the parish of Checkley four miles south-east of Cheadle. The sandstone quarries produce good building material. Whetstones are made here. (pp. 57, 89.)

Ilam (175) is a pretty little village near the junction of the Manifold and Dove and close to the entrance to Dovedale. The

Ilam

church contains a white marble monument by Chantrey and several fine tombs, including the shrine and tomb of St Bertram. Two Saxon crosses stand in the churchyard. The joint waters of the Manifold and Hamps come to the surface in the grounds of Ilam Hall after their underground journey from Darfur Crags. This part of the grounds is so beautiful that it is locally termed " Paradise." " St Bertram's Well " is a spring in one of the lawns. (pp. 13, 20, 69, 74.)

Keele Hall

Ipstones (1482), 4½ miles north of Cheadle. The village is the largest in the Moorlands and stands high with splendid views in the neighbourhood. Iron ore used to be mined.

Keele (1156), near Newcastle, is a village with Keele Hall, an Elizabethan mansion, near by. There is no trace of the building formerly used by the Knights Templars and, later, by the Knights of St John of Jerusalem. (p. 93.)

Kidsgrove (9012), near Stoke, has collieries and engineering works.

King's Bromley (602) lies five miles north of Lichfield. The name means " broom lands," and the district was formerly a royal manor. Leofric, Earl of Mercia, husband of Godiva, had a house here. The church contains early Norman rubble work. (p. 65.)

Kingswinford (20,803) is a "Black Country" town $3\frac{1}{2}$ miles north-west of Stourbridge with coal and iron mines, and iron and brick works. (pp. 52, 58, 112.)

Kinver (2348), an ancient borough on the river Stour, was a resort of the Mercian Kings and gave its name to a large forest which formerly existed here. On Kinver Edge is a large camp, probably prehistoric. Half-way up the hill is "Holy Austin Rock," a sandstone rock carved out into rooms with doors and windows. It is used as a dwelling, though similar places on the other side of Kinver are uninhabited.

Lapley (788), $3\frac{1}{2}$ miles south-west of Penkridge. The interesting church is the only remnant of a priory founded in 1061. Lapley House was captured by the Royalists in the Civil War.

Leek (16,663). "The Capital of the Moorlands" is a centre of the silk industry, and has a large market. There are four Saxon crosses in the churchyard of the parish church, which was dedicated to Edward the Confessor. Charles Edward passed through Leek on his march to Derby and again on his return. Near Leek are the remains of the Cistercian Dieu-la-cresse[1] (or Dieulacres) Abbey. The Roches, north of Leek, form a picturesque and rocky millstone-grit escarpment rising to a height of 1500 feet. Its French name (sometimes mis-spelt Roaches) and the street called "Petite France," call to mind the colony of French prisoners who were lodged in the "barracks" on the Ashbourne road during the Peninsula War, many of whom settled in Leek and the neighbourhood. There s a similar

[1] I.e. God prosper it.

rocky outcrop in Cornwall near a hamlet known as Roche. (pp. 51, 58, 68, 69, 88, 95, 112.)

Lichfield (8616) is an ancient city and municipal borough, and forms a county of itself.

The history of the town is that of its cathedral. At the end of the seventh century it was for a few years the see of an Archbishop with supremacy over Canterbury. The present cathedral was built between 1200 and 1340 A.D. The Close, which had been fortified by Bishop Walter de Langton (1296–1321), was garrisoned for the King in 1643, and was captured after a short siege by Sir John Gell. It was again taken and retaken during the war. Bishop John Hacket (1661–1670) restored the building, but it suffered from neglect during the dark period of the English Church in the eighteenth and early nineteenth centuries. Recently it has been completely restored under the direction of Sir Gilbert and Mr J. O. Scott. There are statues of George Herbert, Izaak Walton, and Dr Johnson on the walls, and here too is Chantrey's celebrated sculpture, the recumbent figures of "The Sleeping Children."

In the Close stand the Bishop's Palace, used by the Sewards when the bishops lived at Eccleshall; the Deanery, Addison's early home; and the homes of Erasmus Darwin, Richard Lovell Edgeworth, and Thomas Day. Dr Johnson's house in Lichfield is now a Johnson Museum; in front of it stands his statue, by Lucas, and close by is a statue of Boswell. David Garrick, Johnson, Ashmole, and Addison attended King Edward VI Grammar School, now removed.

The Whit Monday "Greenhill Bower" is a kind of fair, a survival of the old "Court of array" of the armed citizens. (pp. 2, 52, 63, 64, 67, 72, 78—80, 86, 90, 93, 98, 100, 103, 105, 106, 111, 112, 116, 118.

Longton (37,479), at the south of the Potteries, was called Lane End till 1865, when it was incorporated by Royal Charter.

It now includes the districts of East Vale, Dresden, Florence, and Normacot. The industries are the manufacture of china, earthenware, and bricks, and mining for coal and iron. (pp. 56, 62.)

Loxley, 2½ miles from Uttoxeter, a hamlet claiming to be the birthplace of Robin Hood. A horn known as " Robin Hood's

Tom Moore's Cottage, Mayfield

Horn " is preserved at Loxley Hall. Its owners have held the Hall since 1327.

Madeley (2797), eight miles south-east of Crewe, comprises several hamlets. The church, which is Perpendicular, contains the tomb of John Offley, to whom Izaak Walton's *Compleat Angler* is dedicated. (p. 56.)

Mayfield (1211), near Ashbourne, consists of Church, Middle, and Upper Mayfield. The interesting church is Norman and Decorated. Thomas Moore lived at Mayfield Cottage from 1813 to 1817; he wrote *Lalla Rookh* while there, and the poem beginning "Those evening bells," inspired by the sound of the bells of Ashbourne parish church, often called the Cathedral of the Peak. (pp. 58, 97.)

Milton (2580), four miles north-north-east of Stoke, contains the few remaining fragments of Hulton or Hilton Abbey, a Cistercian house founded in 1223. There are aluminium and chemical works here.

Mowcop (2201, ecclesiastical parish), on the hill of the same name on the Cheshire border (the first syllable rhymes with cow). Fine views are to be obtained from the sham ruin on the top, built by the Wilbraham family about 1760. (p. 7.)

Moxley (4208, ecclesiastical parish) is a modern parish four miles south-east of Wolverhampton, dependent upon the local coal mines and ironworks. Casting sand is obtained here.

Newcastle-under-Lyme (20,201), near, but not part of, the Potteries. The manufacture of hats, once the staple trade of the town, is now discontinued, but brewing, malting, fustian cutting, and the manufacture of cotton and paper are still carried on. There is also a factory for army, railway and police clothing. The High School, re-established in 1872, has 155 boys. (pp. 85, 93, 95, 99, 103, 114.)

Newchapel (6605, ecclesiastical parish), a modern colliery village. James Brindley, the famous canal engineer, lies buried in the churchyard.

Norton Canes or **Norton-under-Cannock** (5648), is a colliery village. It is in the Brownhills Urban District. (p. 23.)

Norton-in-the-Moors (5299), one mile from Milton station, north-east of the Potteries, is a village with collieries and ironworks.

Oakamoor (993) is a modern village in a deep valley of the Churnet 9½ miles south-east of Leek. The first Atlantic cable was made in the large copper wire and tube works here. (p. 57.)

Ogley Hay and **Brownhills** were formed into one ecclesiastical parish in 1854. These two villages, with Shire Oak, Norton Canes and Walsall Wood, form the coal-mining Urban District of Brownhills, with the large population of 16,852. (p. 52.)

Old Hill (11,600) is a modern " Black Country " town in the extreme south of Staffordshire, in the Rowley Regis Urban District.

Penkridge (2386), six miles south of Stafford, was important in Saxon times. The church was one of the six collegiate churches of Staffordshire with a dean and four prebendaries. The deanery, from the reign of King John to Edward VI, was in the hands of the Archbishops of Dublin, who always held it themselves. The church is a large building in Early English, Decorated and Perpendicular styles. The moated Pillaton Hall, two miles away, is partially ruined. (pp. 52, 80, 92, 103.)

Quarry Bank (7393) is a " Black Country " town near Brierley Hill.

The Ridwares, which lie east of Rugeley between the Trent and Blythe, are Pipe Ridware and Mavesyn Ridware, which take their names from the families of Pipe and Mavesyn, and Hamstall Ridware. Hamstall means homestead, and Ridware is derived from rhyd-ware—" river-folk."

Rowley Regis (37,000) is a town south-east of Dudley, Rowley rag, a hard basalt, is quarried here and used for paving and channelling. The industries are mining for coal and iron-stone; and the manufacture of nails, anchors, chains, rivets, gas tubing, gun-barrels, hinges, agricultural implements, pottery, and bricks. (p. 55.)

Rudyard (81), two miles north-west of Leek, is the resort of excursionists who come to see the pretty Rudyard lake, an artificial reservoir for the Trent and Mersey canal. There is good boating and fishing, and the moorlands of north-east Staffordshire are near.

Font in St Mary's Church, Stafford

Rugeley (4504), an old market town on the north of Cannock Chase. The main line of the London and North Western Railway runs through it, as do the Trent and the Trent and Mersey Canals. The water supply comes from bore holes at Fairoak. (pp. 98, 99.)

Salt (400), 3½ miles north-east of Stafford, derives its name from disused salt pits.

Sedgley with **Upper** and **Lower Gornal** (16,527) are "Black Country" parishes two or three miles south of Wolver-hampton. There is much fire-clay and coal in the district, and there are ironworks. (p. 53.)

Silverdale (7795) near Newcastle may have once deserved its name, but is now principally a coal-mining and iron-working district. It is governed by the Wolstanton United Urban District Council.

Smallthorne (13,559), a pottery town adjoining Burslem.

Smethwick (70,694) is a large manufacturing town three miles north-west of Birmingham, of which it practically forms a suburb. (pp. 53, 57, 58, 111.)

Stafford (23,383), the county town, situated on the river Sow almost in the centre of the county. St Bertalin was a hermit who lived on an island (Bethney, now called Stafford) early in the eighth century. A church in Stafford, dedicated to him, remained until the eighteenth century. The site has thus been occupied for 1200 years. In 1206 King John gave the town a charter, and Edward VI and James I granted various rights and privileges. Elizabeth visited it, and, after hearing complaints of the burghers, protected the staple industry of cap-making by her grant of the "Statute of Capping." Two members were returned to Parliament from 1295 to 1885, when the number was reduced to one. Sheridan was member from 1780 to 1806.

St Mary's church was a collegiate establishment with thir-teen prebendary canons at the time of Domesday Book, 1086. It was also a Royal Free Chapel. It is built in the Early English and Decorated styles with work of later types. Izaak Walton was born in the parish and baptized in the Norman font. St Chad's is, next to Tutbury, the finest Norman church of the county. On the tower is the inscription ORM VOCATUR QUI

ME CONDIDIT.   Orm was a not uncommon Norman name.   Much
of the edifice has been rebuilt and restored at various times.
There are many interesting houses in the town.   Stafford is an
important railway centre.   The chief industry is the making of
boots and shoes, but there are also salt works and engineering
works.   (pp. 2, 58, 65, 67, 68, 75, 78, 82, 84, 86, 88, 90, 92,
93, 96, 99, 102, 103, 105, 106, 112, 115.)

Council Chamber, Stoke

Stoke.   The six towns which were federated in 1910 to
form the one town of Stoke-upon-Trent were Burslem (41,566),
Fenton (25,626), Hanley (66,255), Longton (37,479), Stoke
(36,375), Tunstall and Goldenhill (27,390) giving a total population of 234,691.   These form the Potteries.   Stoke and Burslem
are old towns, but the others are growths of the nineteenth century.   Stoke is an important railway centre, and here are the

S. S.   10 works of the North Staffordshire Railway Company. (pp. 21, 62, 111, 112, 118, 122.)

Stone (5688) is an ancient market town seven miles north of Stafford. It probably takes its name from some vanished monument. Bury Bank is a camp near. The name Wulfherecester was given to it because Wulfhere, King of Mercia, occupied it for some time. The Duke of Cumberland encamped in Stonefield, a suburb, in 1745. The ruins of the abbey of Wulfad and Rufin, sons of Wulfhere, are at the south of the town and there

The Chancel, Tamworth Church

is a Dominican Convent in it. Admiral Sir John Jervis, afterwards Earl of St Vincent, was born near here in 1735. (pp. 21, 58, 71, 90, 99, 103.)

Swythamley, seven miles from Rushton, is near the rocky chasm known as Ludchurch, at the north-west end of the Roches. It was used by the Lollards as a secret place of worship.

Talke (5497), commonly called Talk-o'-th' Hill, though the prefix means a hill, is a mining village on the north-west border five miles north-west of Newcastle. It has suffered serious calamities at various times from colliery explosions, fires, and a great gunpowder explosion. (p. 99.)

Tamworth (7738), the "worth" or farm on the Tame, is a very old town. Part of the ecclesiastical parish is in Warwickshire, as was part of the town till 1890, when for local government purposes it was added to Staffordshire. Offa, King of Mercia, built a famous house here about 790 and dug around the town "Offa's Dyke" or the "King's Ditch," of which portions can still be seen. Destroyed by the Danes in 874, the town was fortified by Ethelfleda in 913. Coins were minted here from the time of Offa till Henry I. The church, of the Decorated style, dedicated to St Editha, daughter of Athelstan, is large and handsome. The soil near the town is very rich. There are valuable mines of coal, fire-clay, and blue and red brick-clay, as well as some factories and paper-mills. (pp. 2, 8, 23, 52, 58, 65 67, 72, 75, 82, 88, 101, 103, 112.)

Tettenhall (5381) is a suburb of Wolverhampton two miles north-west of the town. Edward the Elder defeated the Danes here in 910. The church of St Michael and All Angels was a collegiate establishment until the Reformation. It contains some Norman work and miserere seats. Wrottesley, close by, gives its name to a well-known Staffordshire family. (pp. 48, 65, 92, 112.)

Tipton (31,756), is a "Black Country" town close to Dudley engaged in the iron trade, particularly of heavy articles such as forgings, boilers, anchors, and railway goods, though smaller articles are made in great variety. Galvanising is carried on. There are cement and brick-works and malt-houses. (pp. 53, 55.)

Tividale, adjoining Tipton and Rowley Regis, has a similar trade to the former place.

Trentham (3059) is an extensive parish with pleasantly situated village, near Newcastle. The Duke of Sutherland had a fine mansion here, recently pulled down. The grounds are now a public pleasure park. The church, of modern rebuilding, contains some fine tombs. There are no remains of the nunnery founded about 680 A.D. by Ethelred, King of Mercia. (pp. 14, 21, 86, 90, 93.)

Trentham Hall

Tunstall (22,494) is part of Stoke. It has extensive iron-works in addition to its earthenware trade. Coal and ironstone are raised. There are brick and tile works. (p. 62.)

Tutbury (2186), six miles north-west of Burton-on-Trent, has a Norman church; a castle, built before the Conquest, in which Mary Queen of Scots was confined; and glass, gypsum, and alabaster industries. (pp. 58, 65, 67, 76, 78, 84, 93.)

Uttoxeter (5717) (pronounced Utoxeter or Ŭxetter), is supposed to have been the Roman Uxacona. Its charter is dated

1251. From 1320 to 1625 the town was a royal demesne, and was at one time held by John of Gaunt. The church, which was rebuilt in 1828, has a tower and spire dating from the time of Edward II. The Duke of Hamilton's army surrendered here in 1648. Dr Johnson did penance in the market-place. The making of agricultural implements is here an important industry. (pp. 52, 63, 68, 72, 86, 100, 102, 117.)

Walsall (92,115) includes the eight wards of Birchills (13,901), Bloxwich (8411), Bridge (8215), Caldmore (16,463), Hatherton (9165), Leamore (11,286), Paddock (10,429), and Pleck (14,245). The old town was a borough by prescription, and received confirmatory charters from Edward III and Henry IV. Another charter was granted by Charles I and confirmed by Charles II. The town manufactures harness, saddlery and saddlers' ironmongery. (pp. 5, 52, 53, 55, 103, 111, 112, 124.)

Wednesbury (28,103), popularly called "Wedgebury," is a "Black Country" town. Its Saxon name was most likely "Woden's Bury." St Bartholomew's church is nearly all modern, with some beautiful modern glass. The iron manufactures are very important. They include rails, boiler plates, bar iron, Bessemer and Siemens-Martin steel, and every sort of wrought ironwork. There are stoneware potteries and mines for coal, iron-ore, limestone, potter's and brick clay. (pp. 23, 52, 53, 55, 112.)

Wednesfield (6488), two miles north-east of Wolverhampton, has extensive factories of locks and keys. (p. 55.)

West Bromwich (68,332), anciently the seat of a Benedictine Priory, is a "Black Country" town of quite recent growth, manufacturing a great variety of small metal goods such as fireirons, locks, coach furniture, spring balances, horse-shoes, saucepan handles, spades, etc. There are iron-smelting furnaces, foundries and forges; brass foundries, boat yards, lime-kilns, tar distilleries, chemical works, etc. (pp. 53, 55, 58, 88, 111, 112.)

Norman Cross, St Peter's, Wolverhampton

Willenhall (18,844) lies three miles east of Wolverhampton. The manufactures include all varieties of locks, bolts, latches, keys, gridirons, hinges, ferrules and other small metal goods. (p. 55.)

Wolstanton (12,395), "Wolstan's Town," is a small town of the Pottery district. It is said to be of Saxon origin. The old parish has been divided into several ecclesiastical districts, namely Newchapel, Chesterton, Knutton, Mowcop, Silverdale, Tunstall, Golden Hill and Kidshead. The parishes of Wolstanton, Chesterton and Silverdale were in 1904 formed into a United Urban District, population 27,335 in 1911. (p. 48.)

Wolverhampton (95,328) is the largest town of the "Black Country." The old town was called Hauton, Heauton, or Hamton, the "town on a hill." In 996 Wulfrana or Wulfrun, sister to King Edgar, founded a college here with dean and prebendaries, and gave it many privileges. The old parish church is mainly in the Perpendicular style of about 1480. It is very large, and cruciform in shape, and has an ancient stone pulpit. The other churches are modern. There is at St Peter's an interesting Norman cross, 12 feet in height which is of late twelfth century date.

Wolverhampton School, founded in 1515, was re-organised as a Public School in 1874; large extensions have been made in the buildings in recent years, and at present (1915) the number of boys is 320.

Remains of Roman iron furnaces have been discovered in the district. The present manufactures include ironmongery and hardware, locks and keys, "steel toys" (a name given to such articles as corkscrews, snuffers, nutcrackers and nippers), bicycles, electric light apparatus, safes, iron and tin japanned goods, papier-maché articles, tea-trays and shoe-tips. There are iron and brass foundries, chemical works, coal mines, ironstone mines, and a great variety of other industries. (pp. 5, 8, 22, 28, 52, 53, 55, 80, 92, 98, 99, 100, 103, 111, 112, 114.)

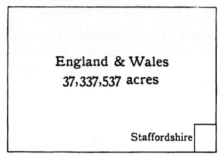

Fig. 1.   Area of Staffordshire (741,320 acres) compared
with that of England and Wales

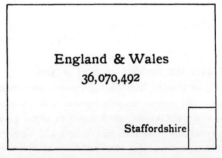

Fig. 2.   The population of Staffordshire (1,348 259) compared
with that of England and Wales in 1911

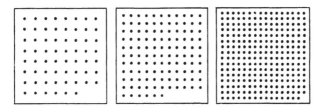

England and Wales 618   Staffordshire 1164   Lancashire 2550

Fig. 3. Comparative Density of Population to the
square mile in 1911

(*Each dot represents ten persons*)

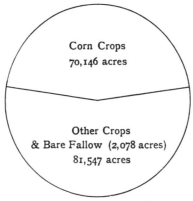

Fig. 4. Proportionate area under Corn Crops compared with
that of other cultivated land in Staffordshire in 1913

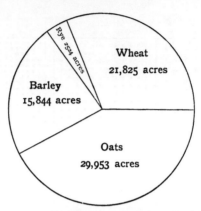

Fig. 5.   Proportionate area of chief Cereals in
Staffordshire in 1913

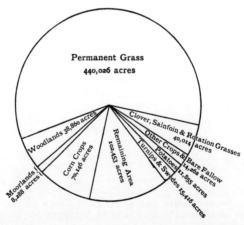

Fig. 6.   Proportionate areas of land in Staffordshire in 1913

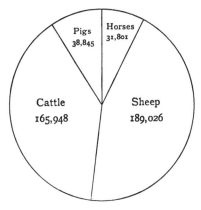

Fig. 7. Proportionate numbers of Horses, Cattle, Sheep
and Pigs in Staffordshire in 1913

www.ingramcontent.com/pod-product-compliance
Ingram Content Group UK Ltd.
Pitfield, Milton Keynes, MK11 3LW, UK
UKHW042144280225
455719UK00001B/102